Lithium in Psychiatric Therapy: Celebrating 75th Anniversary

Lithium in Psychiatric Therapy: Celebrating 75th Anniversary

Guest Editors

Janusz K. Rybakowski
Ewa Ferensztajn-Rochowiak

MDPI

Basel • Beijing • Wuhan • Barcelona • Belgrade • Novi Sad • Cluj • Manchester

Guest Editors

Janusz K. Rybakowski
Department of Adult Psychiatry
Poznan University of Medical Sciences
Poznan
Poland

Ewa Ferensztajn-Rochowiak
Department of Adult Psychiatry
Poznan University of Medical Sciences
Poznan
Poland

Editorial Office

MDPI AG
Grosspeteranlage 5
4052 Basel, Switzerland

This is a reprint of the Special Issue, published open access by the journal *Pharmaceuticals* (ISSN 1424-8247), freely accessible at: https://www.mdpi.com/journal/pharmaceuticals/special_issues/3K5B15GY40.

For citation purposes, cite each article independently as indicated on the article page online and as indicated below:

Lastname, A.A.; Lastname, B.B. Article Title. *Journal Name* **Year**, *Volume Number*, Page Range.

ISBN 978-3-7258-7605-1 (Hbk)
ISBN 978-3-7258-7606-8 (PDF)
https://doi.org/10.3390/books978-3-7258-7606-8

Contents

About the Editors

Janusz K. Rybakowski

Janusz K. Rybakowski, MD, PhD, is presently Emeritus Professor at the Department of Adult Psychiatry, Poznan University of Medical Sciences, Poznan, Poland. He was the Head of this department from 1995 to 2016 and the Chairman of the Department of Psychiatry at the Medical Academy in Bydgoszcz from 1985 to 1995. In 1976–1977, he was an NIH Fogarty Research Fellow in the Department of Psychiatry at the University of Pennsylvania, Philadelphia. Prof. Rybakowski was the President of the Polish Psychiatric Association from 1998 to 2001, a Board Member of the Association of European Psychiatrists from 1998 to 2004, and a member of ten other international scientific organizations. In 2021, he became a corresponding member of the Polish Academy of Sciences, the only Polish psychiatrist to be elected to the Academy in its 70-year history. In 2025, he was promoted to full membership of the Academy. He authored over 700 scientific articles on clinical psychiatry, neurobiology, and psychopharmacology. His special clinical and research interest is lithium treatment of mood disorders. He is the author of the books "The Faces of Manic-Depressive Illness" (2009), "Genetic Influences in Response to Drug Treatment in Major Psychiatric Disorders (with Alessandro Serretti, 2016), and "Lithium – the Amazing Drug in Psychiatry" (2020). He is the Board President of the journal *Psychiatria Polska*, Editor-in-Chief of *Pharmacotherapy in Psychiatry and Neurology*, and *Neuropsychiatry and Neuropsychology*, and a member of the editorial boards of twelve international journals. In 2012, received the Lifetime Achievement Award from the European Bipolar Forum; in 2015, the Lifetime Achievement Award in Biological Psychiatry from the World Federation of the Societies of Biological Psychiatry; and in 2018, the Mogens Schou Scientific Award from the International Society of Bipolar Disorders. He is the only Polish psychiatrist in the Elsevier-Stanford University ranking of the top 2% of world scientists.

Ewa Ferensztajn-Rochowiak

Ewa Ferensztajn-Rochowiak, MD, PhD, is a psychiatrist and assistant professor in the Department of Adult Psychiatry at Poznan University of Medical Sciences, Poland, where she heads the Affective Disorders Laboratory. Her clinical and research interests focus on bipolar disorder and depression, with an emphasis on lithium treatment, immunopsychiatry, clinical staging, and personalized medicine. She is actively involved in numerous national and international research projects, including studies on lithium prophylaxis, sensory processing sensitivity, treatment-resistant depression, and predictors of mood disorders' clinical course. She coordinated the international PROMPT project under ERAPerMed and is a member of the International Group for the Study of Lithium-Treated Patients (IGSLI). Her achievements include a doctoral degree in familial bipolar disorder and a postdoctoral degree focused on stem cells in bipolar disorder. She is an international consultant in positive psychotherapy (World Association for Positive and Transcultural Psychotherapy, WAPP) and has received multiple awards from the Rector of the Poznan University of Medical Sciences in recognition of her scientific contributions.

Preface

This Reprint details the prominent role of lithium as a psychopharmacological drug after the three-quarters of a century since its introduction to modern psychiatry. Ten articles by leading specialists on lithium discuss the development and current status of lithium treatment. Lithium is now the first drug of choice for the prevention of mood recurrences in bipolar disorder. This Reprint outlines the current views on lithium's use, along with its mechanisms, treatment optimization, and some controversies. Among mood-stabilizing medications, lithium has a conspicuous anti-suicidal activity, and the work in this Reprint covers the history of this effect and recommendations for its application. In recent decades, researchers have also discovered neuroprotective effects of lithium, making it a candidate for use in the prevention and treatment of dementia disorders, which is also explored here. The Reprint is targeted at the psychiatric community and aims to broaden their knowledge about this extraordinary psychotropic drug.

Janusz K. Rybakowski and Ewa Ferensztajn-Rochowiak
Guest Editors

pharmaceuticals

MDPI

Review

The Rise of a Legend: Lithium and the Extraordinary Story of Its Discovery

Konstantinos N. Fountoulakis [1,2]

[1] 3rd Department of Psychiatry, School of Medicine, Faculty of Health Sciences, Aristotle University of Thessaloniki, 54124 Thessaloniki, Greece; fountoul@auth.gr or kostasfountoulakis@gmail.com

[2] Society for Neurosciences and Rehabilitation (E.N.A.), 54124 Thessaloniki, Greece

Abstract: Lithium is still the gold standard in the treatment of Bipolar disorder in adults. Despite the recent advances and the continuous increase in the use of antipsychotics and specific antiepileptics, it is still widely used despite not being promoted by the industry. Lithium was discovered in the early 19th century in a Swedish mine, and only a few decades later, it was discovered that it could be added to water so that urea could be dissolved with the formation of lithium urate. This was considered in the frame of the uric acid diathesis theory, which, during the 19th century, was considered a strong etiopathogenetic theory for many pathological conditions, including mental disorders. The first attempts to use lithium as a treatment modality in psychiatry appeared in the late 19th century by Hammond and the Lange brothers, and during the late 19th century and the early 20th century, lithium was used in various remedies and beverages until its toxicity became evident. Lithium was established as a valid treatment option for Bipolar disorder only during the second half of the 20th century after the works of John Cade, Mogens Schou, Poul Baastrup, and others who pursued a line of research, frequently against the scientific opinion of the majority of colleagues.

Keywords: lithium; history; bipolar disorder; discovery

1. Introduction

Lithium is still the gold standard in the treatment of Bipolar disorder. Despite recent advances in the field and the increased use of antipsychotics and specific antiepileptics, it is still widely used despite the lack of promotion by the industry. Its history is of particular interest, and reviewing it could lead to insights into how our understanding of mental illness evolved and the modern era of psychopharmacology emerged. Lithium carbonate is now on the World Health Organization's list of essential medicines.

The current paper aims to give technical and historical information on this extraordinary and mysterious medication.

2. What Is 'Lithium'?

Lithium is a relatively rare chemical element (16 ppm in the Earth's crust, 0.006 ppm in the universe) [1]; its symbol is 'Li'. It belongs to the alkali metal group (Figure 1); it is the least dense solid element and the lightest metal in the periodic table [2]. Lithium is silver–white and soft enough to cut with a knife. Its standard atomic number is equal to 3, and its atomic weight is equal to 6.941 g·mol^{-1}. Its melting point equals 180.5 °C (356.90 °F), and its boiling point is 1342 °C (2448 °F), and both are among the lowest points for a metal. Due to its low atomic mass, lithium has high electrode potential because it has

https://doi.org/10.3390/ph18081230

a high charge and power-to-weight ratio. Lithium has good heat and electric conductivity in its metallic form, enabling it to store and transmit energy. This makes it well-suited for the building of rechargeable batteries. There are two stable lithium isotopes in nature (Li_6 and Li_7, with the latter accounting for 95.15%).

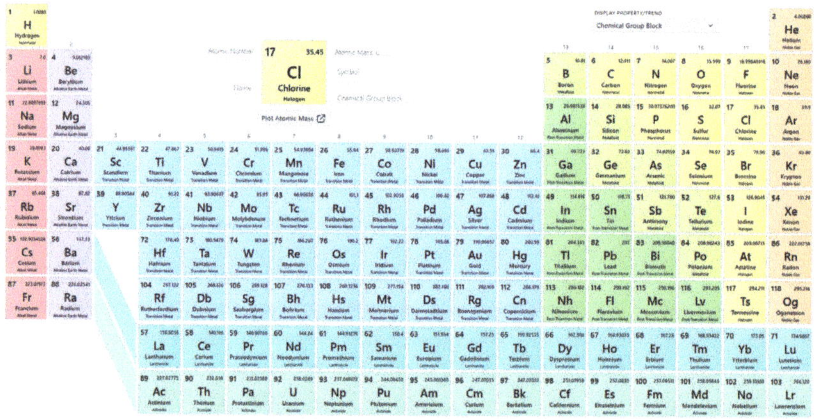

Figure 1. The periodic table, from https://pubchem.ncbi.nlm.nih.gov/periodic-table/ (accessed on 15 August 2025).

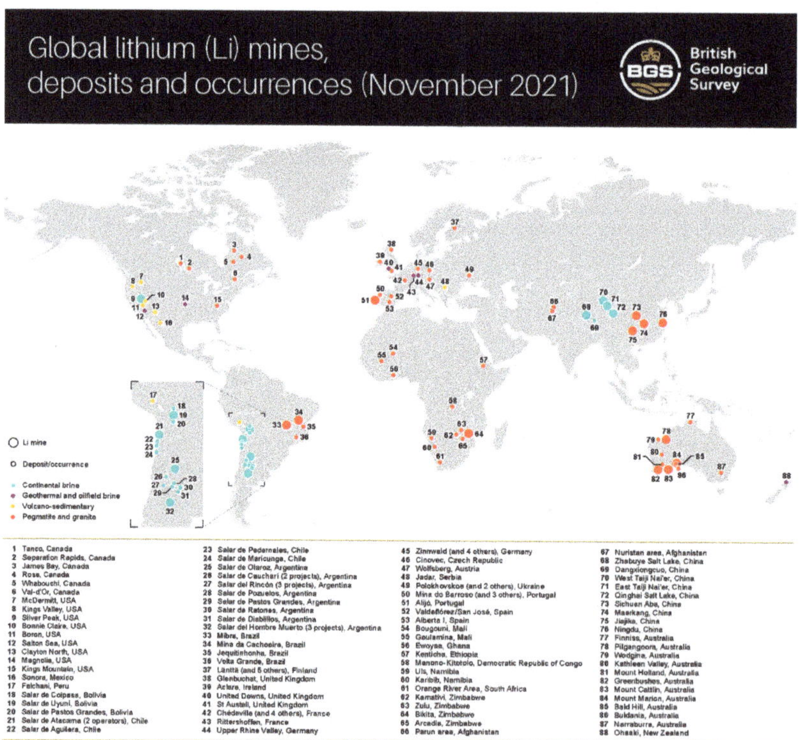

Figure 2. Global lithium mines and deposits according to the British Geological Society. From Shaw, R.A. (2021) Global lithium (Li) mines, deposits and occurrences (November 2021). British Geological Survey.

Lithium metal is highly reactive. It ignites in contact with water. As a result, in nature, it is only found as a mineral or salt and almost never in a pure form. Metallic lithium is soluble in short-chain aliphatic amines, e.g., etilamine. On the contrary, it is insoluble in hydrocarbons. Today, minerals rich in lithium, including spodumene and petalite, are mainly extracted from pegmatites in Australia, Zimbabwe, and Brazil (Figure 2) [3]. Other sources of lithium include oilfield brines, clays, and geothermal brines.

3. The History of Lithium

3.1. The Discovery (First Half of the 19th Century)

Lithium was discovered by José Bonifácio de Andrada e Silva (1763–1838), a Brazilian chemist and statesman, in 1800, in a mine on the island of Utö, Sweden, in the form of petalite ($LiAlSi_4O_{10}$; Figure 3) [4].

Figure 3. Petalite, the mineral where originally lithium was discovered, from Minas Gerais State, Brazil. Photo by Eurico Zimbres. Copyright status: CC BY 2.0.

It was re-discovered in 1817, this time by Eric Thomas Svedenstjerna (1765–1825), a Swedish metallurgist. Svedenstjerna sent samples to the laboratory of Jöns Jacob Berzelius (1779–1848), the foremost Swedish chemist of the day. There, Johan August Arfwedson (1792–1841) analyzed it, and he found that 96% of the petalite consisted of alumina and silica. The remaining 4% was an alkaline substance that did not respond to the usual tests for known alkaline elements. Eventually, Arfwedson identified a new element, and Berzelius, being the head of the laboratory, gave it the name 'lithion' or 'lithina', from the Greek word 'λιθος' ('lithos' meaning "stone"). The name reflected its discovery in a solid mineral. On the contrary, potassium was discovered in plant ashes, while sodium is highly abundant in animal blood [5,6].

Both Arfwedson and Gmelin tried but failed to isolate the pure lithium from its salts. The pure element was isolated in 1821 by William Thomas Brande (1788–1866) by electrolysis of lithium oxide. In 1855, Robert Bunsen (1811–1899) and Augustus Matthiessen (1831–1870) produced larger quantities of lithium through the electrolysis of lithium chloride [7]. After the discovery of this procedure, the German company Metallgesellschaft AG started the commercial production of lithium in 1923 [8].

3.2. The 'Uric Acid Diathesis' Theory (Second Half of the 19th Century)

Already in the mid-nineteenth century, several medical practitioners noted that lithium compounds could dissolve uric acid. Lithium combines with uric acid to form lithium urate ($C_5H_3LiN_4O_3$), a soluble salt. At the time, prominent doctors believed that excess uric acid was the culprit behind many severe illnesses. So, based on its ability to dissolve uric acid, physicians of the mid-19th century began treating many diseases, including asthma, diabetes, obesity, indigestion, rheumatism, skin disorders, and headaches, with lithium.

In 1841, Alexander Lipowitz (1811–1873), a German chemist, pulverized some lepidolite, an ore containing lithium, mixed it in boiling water with barely soluble uric acid, and found that he had produced the far more soluble lithium urate. Two years later, Alexander Ure (1808–1866), a British surgeon and father of the field of drug metabolism [9], during a lecture to the Pharmaceutical Society, showed how lithium carbonate, when put together with uric acid crystals, formed lithium urate. This mechanism dissolves uric acid crystals. Ure had a particular interest in finding a remedy for gout and for the urinary stones that often accompany it, at least partially because his father, an eminent scientist as well, had suffered from this disease for many years. His observation suggested that injecting lithium carbonate directly into the bladder might be a good way of reducing the size of bladder stones [10]. It took 16 years, but finally, in 1859, Ure was able to get his hands on enough lithium carbonate to try it as a treatment on a 56-year-old male patient with a large and uncomfortable bladder stone. He injected a solution of lithium carbonate directly into the bladder. He did this every other day for several weeks. Disappointingly, the stone did not get any smaller. To 'expedite matters', Ure resorted to crushing the stone with a lithotrite, an instrument inserted into the bladder. After performing this several times, he had broken up much of the stone and a successful conclusion seemed likely. Unfortunately, before Ure removed the final fragments of stone, the patient died. In his report of the case, Ure wrote that at least 'the solvent had in some measure lessened the cohesion of the concretion and increased its friability' [11].

Sir Alfred Baring Garrod (1819–1907) was a prominent British physician who announced that he had discovered uric acid in the blood of gouty patients. He was a contemporary of Ure and was familiar with Ure's bladder stone work. He repeated his original experiment with a gouty bone instead of a bladder stone. He placed a finger bone, which featured a prominent gouty (uric acid) deposit at the joint, in a solution of lithium carbonate. Within several days, the deposit had disappeared. Garrod propagated the notion that excess uric acid was behind an assortment of other medical conditions, from headaches to diabetes, also including many mental disturbances.

It is interesting that according to the works of Armand Trousseau (1801–1867) in France and Alexander Haig (1853–1924) in the UK, mania and depression were also thought to be related to uric acid (brain gout). Being a prominent Scottish physician, Alexander Haig is said to have cured his migraines by switching to a vegetarian diet, which is low in uric acid. Subsequently, based on his personal experience, Haig concluded that behind all illnesses, there was a uric acid disorder that served as the cause. He described it as 'uric acid diathesis' (a tendency to produce too much uric acid) [12,13]. This concept, the 'uric acid diathesis' theory, captivated the medical profession. Garrod advised that lithium carbonate be taken orally as a treatment for gout as well as for other uric acid diathesis conditions. By the end of the 19th century, influential physicians throughout Europe and the US had embraced the concept and strongly recommended lithium for a variety of conditions. Unfortunately, eventually, it was proven that in order to dissolve urate in the body, the lithium levels needed would be toxic [14–17]. The uric acid diathesis remained a leading theory about the cause of disease well into the 20th century.

Garrod and other advocates of the uric acid diathesis believed that mental symptoms commonly went along with gout and excess uric acid. They thought that uric acid attacked the brain and, as a result, caused all sorts of mental disorders, including melancholy and, as Garrod put it, 'gouty mania'. Garrod and others treated all signs of uric acid diathesis with lithium compounds and reported that, along with other symptoms, mental ones improved as well.

The first neuropsychiatric indication for lithium came from the eminent American neurologist Silas Weir Mitchell (1829–1914). In 1870, Mitchell recommended lithium bromide as an anticonvulsant and hypnotic [18] and later for 'general nervousness' [19]. It worked fairly well, probably because bromine is a good sedative, and it continued as an epilepsy treatment well into the 20th century. Mitchell also gave lithium bromide to people with an assortment of nervous symptoms, some of whom were probably suffering from depression, and reported that they received relief. Again, the relief may have come from the bromine portion of the compound, not the lithium.

Then, in 1871, William Alexander Hammond (1828–1900; Figure 4) was maybe the first to prescribe a modern and effective psychotropic agent specifically for a mental disorder, and this was lithium for acute mania [20]. At that time, he was a professor of Diseases of the Mind and Nervous System at the Bellevue Hospital Medical College in New York. In his 1871 textbook on diseases of the nervous system, Hammond reported that he had used lithium bromide to treat patients with acute mania and discovered that it rapidly and effectively calmed them. He used exceptionally high doses. It is not clear whether the lithium or the bromine part produced the results. Additionally, it was unclear if some kind of sedation due to toxicity accounted for this. In his later textbooks, published in 1882, 1883, and 1890, Hammond did not mention the use of lithium, suggestive of a disappointment [21,22]

Figure 4. William Alexander Hammond (1828–1900).

A few years after Hammond, the Danish psychiatrist Carl Lange (1834–1900; Figure 5) used lithium in the treatment of recurrent brief depression in 1886, while his brother Frederik Lange (1842–1907) used lithium in the treatment of 35 patients with melancholic depression (including some milder forms of Bipolar disorder) in 1894 [23]. In 1886, the year after he published his work on emotion, he came out with a treatise on what he called periodical depressions. In it, he described a series of patients he had cared for in his medical practice who had repeated depressions of milder severity than is usually seen by psychiatrists. He noticed what he thought was an unusual sediment in these patients' urine and assumed, incorrectly, that the sediment was uric acid. Lange went on to hypothesize that these depressions were caused by excess uric acid and suggested that they could be treated and prevented by the conventional (at the time) therapies for uric acid conditions, which included lithium carbonate. In later writings, he explicitly mentioned lithium carbonate as a treatment for these 'uric acid depressions'. The lithium doses he advised were comparable to those used today for Bipolar disorder, and Lange reported favorable outcomes with this regimen. Lange's colleagues in Denmark knew of his work, but for the most part, the larger international psychiatric community did not.

Figure 5. Carl Georg Lange (1834–1900).

Eventually, despite encouraging results, by the turn of the 20th century, the 'brain gout' theory of mood disorders disappeared as a medical concept, and the use of lithium in psychiatry was abandoned.

3.3. *Remedies and Beverages (Late 19th Century and First Half of 20th Century)*

However, while the scientific medical community rejected the gout theory, Haig's writings were extremely popular with a large percentage of the lay public. With the advent of uric acid diathesis as the explanation for an assortment of illnesses and the widespread application of lithium as its treatment, the promotion of mineral waters now had a 'scientific rationale'. By the last decade of the 19th century, people were flocking to mineral springs and the resorts that sprang up around them. Some of these springs contained trace amounts of lithium and some no lithium at all, but these 'lithium waters' were sought as a cure for rheumatism, diabetes, asthma, gout, and every other imaginable ailment. People also prized them for what were widely believed to be their general health-promoting properties, and both bathed in and drank the waters. 'Lithia water' became widely available to the general public without recommendation or prescription. Innumerable brands of bottled lithium water came on the market, each promising extraordinary health benefits. Multiple companies sold water labeled as containing lithium, one of which continues to operate as a business presently.

Also, several beverages included lithium as a component in the late 19th and early 20th centuries (Figure 6). Charles Leiper Grigg (1868–1940) introduced a lemon–lime soft drink in 1929 under the label 'Bib-Label Lithiated Lemon-Lime Soda', which soon changed to 7UP. Lithium citrate was one of its seven main ingredients and a key selling point. Soon after launch, its inventor changed the name to 7Up Lithiated Lemon Soda, and in 1936, he shortened the name to 7 Up. The lithium was not removed until 1950. Also, Coca-Cola offered a lithium-containing version that did not last long. Eaux de Vichy is another example of natural water advertised as containing lithium.

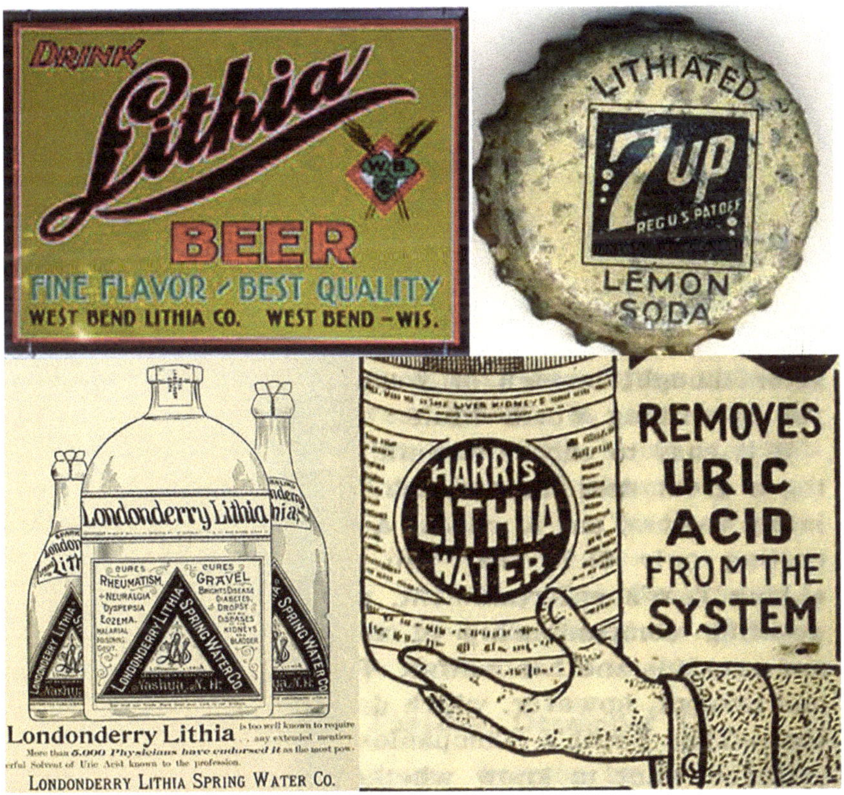

Figure 6. Examples of beverages which included lithium as a component in the late 19th and early 20th centuries. Left upper corner: Lithium beer. Right upper corner: The original 7UP which included lithium. Bottom: An example of remedies with lithium which were marketed in the late 19th and early 20th centuries and were mostly indicated for the control of renal calculi and 'uric acid diathesis'.

Later studies eventually showed that lithium taken by mouth has no effect on uric acid or its elimination, partly because other compounds circulating in the body interfere with lithium's dissolving ability. By 1948, lithium had been removed from all beverages because of cases of toxicity, and its free marketing was prohibited [14].

3.4. The Second Half of the 20th Century and the Re-Discovery of Lithium

After World War II (WWII), a boom in biomedical research appeared that eventually led to the emergence of the modern era of psychopharmacology. Although this era of psychopharmacology is considered to have begun in the 1950s with the introduction of antipsychotics, in reality, it started a decade earlier with the re-discovery of lithium by John Cade (1912–1980; Figure 7) in 1949 [15].

Figure 7. John Cade (1912–1980) from the Herald & Weekly Times Portrait Collection, State Library of Victoria, Australia.

John Cade was born on 18 August 1912, in a country town called Murtoa near Melbourne in Australia. Cade's father, David Cade, worked there as a General Practicioner, and his mother worked there as a nurse. Not psychologically able to deal with the demands of his position due to Post-Traumatic Stress Disorder (PTSD) from World War I (WWI) and the Spanish flu pandemic, David Cade sold his private practice and started working as an asylum doctor in Victoria's Mental Hygiene Service. For the next decades, he served as the medical superintendent at several Victorian mental hospitals. An interesting detail is that at that time, it was customary for a house within the grounds of the asylum to be provided to the superintendent and his family. This led John Cade to develop a deep understanding of the needs and the burden of mental patients and their families [24].

By 1936, at the age of 24, Cade was appointed to Beechworth Mental Hospital, the same hospital where his father had worked and the Cade family had lived 15 years earlier. Three years later, he moved to the department's Bundoora Repatriation Mental Hospital. There, he had an exceptional experience in research since, during a flu outbreak, he worked with Sir Frank Macfarlane Burnet (1899–1985), an eminent virologist and immunologist. Burnet received the Nobel prize in 1960. The research project concerned measuring antibodies in the blood of afflicted patients [25].

With the break of WWII, Cade joined the army, and in September 1941, he was promoted to major. Then things went badly as Singapore fell to the Japanese on February 15th, 1942, and Cade was taken as a prisoner of war along with around 80,000 RANZAC soldiers. Cade was put in charge of the psychiatric section in a hospital organized in the Changi prisoner of war camp, and there he began to conceptualize a relationship between diseases and food deficiencies in a way parallel to that of Archie Cochrane in the frontstalage 183 German prisoners of war camp in Thessaloniki, Greece. During that period, he also developed the solid idea that severe mental disorders have a physical basis.

A second critical figure in the re-discovery of lithium was Eduard (later known as Edward) Trautner (1890–1978; Figure 8) [26]. Trautner was of German origin and was a scientific polymath; he studied several sciences, including animal physiology, botany, and biochemistry. He arrived in Australia on the HMT Dunera, a ship that achieved notoriety for the appalling conditions of its voyage to Sydney, in early September 1940 with another 2542 so-called enemy aliens who were deported to Australia by the British government Most of them were Jewish refugees, and others who opposed the Nazi regime and had fled Germany. After arriving in Sydney, he was sent to an internment camp in Tatura, central Victoria, where he was discovered by the eminent professor of physiology at the University of Melbourne, Sir (AK) Roy Douglas 'Pansy' Wright (1907–1990). Wright arranged for Trautner to join Melbourne's physiology department in 1942. Trautner worked there for the next few years, developing procedures aimed at extracting pharmaceutical agents from plants.

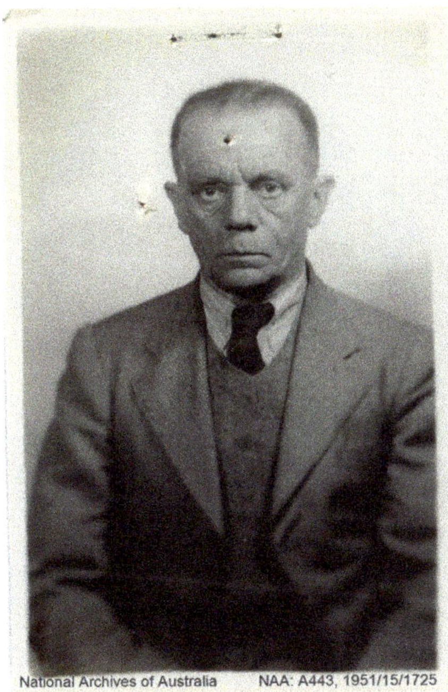

Figure 8. Eduard (later known as Edward) Trautner (1890–1978).

After the war, John Cade started working on research using abandoned spaces in Bundoora Repatriation Mental Hospital near Melbourne. Cade had little formal research training, and he went on to create a makeshift laboratory on the grounds of a small, isolated mental asylum in an unused kitchen. He had no access to sophisticated chemical analysis and no theory to guide him. He had neither funding nor collaborators.

His belief that there is a physical dysfunction underneath mental disorders led him to the assumption that severe mental disorders occur because the body produces a high amount of an otherwise normal substance, and he formulated a theory in analogy with hyperthyroidism, for whom medication treatment had started emerging around that time. According to this conceptualization, schizophrenia and mania were the result of the excess of this substance, while melancholia was its deprecative condition.

Since he had no idea what the substance might be, he thought that the best way to identify it would be by studying conditions like mania and psychosis, where this substance is supposed to be an excess. In this line of thought, if such a substance does exist in excess, then either itself or one of its metabolites would be present in the urine in excess. With this in mind, Cade started collecting early-morning urine samples from patients as well as from healthy persons because these urine samples were quite concentrated since they were passed after 12–14 h without fluid intake. Theoretically, the suspected toxic substance would exist in its highest concentration in these samples. The next step was again a step into the unknown. The substance was unknown, and its physiological actions were also unknown. If it was toxic when in excess, then some kind of toxicity test could reveal it. Cade decided to go on with a very crude toxicity test, and in 1949, he simply injected various amounts of urine into the abdominal cavities of guinea pigs. He titrated the dosage and determined the quantity required to kill them. He observed (mistakenly) that the urine of people with mania was especially lethal to the animals. He registered that the urine from manic patients would kill guinea pigs in a dose as low as 0.25 mL per 30 g of body weight, whereas the lethal dose of the most toxic urine from patients with other diagnoses and healthy people was at least 0.75 mL per 30 g of body weight. However, the mode of death was the same, regardless of the source of the urine, pointing to the possible conclusion that there was an issue of quantity, not quality. Twenty minutes after injection, the animal started to tremble and lose its balance; convulsions followed, leading to coma and death in a state of status epilepticus. To identify the toxic substance, Cade embarked on a series of experiments without following an entirely logical or readily understood sequence; most likely, they were a kind of fishing expedition. Even 'worse', in the process, Cade misinterpreted critical aspects of the observations.

His first step was to test the toxicity of the common contents of urine, which are urea, uric acid, and creatinine, which are all normal products of protein metabolism. After injecting high concentrations of each of them into the abdominal cavities of guinea pigs, he observed that creatinine and uric acid were harmless, but on the contrary, urea killed the animals in the same manner urine did. This could mean that urea was the substance responsible for the lethal effect of urine. The problem was that there was no higher concentration of urea in the urine of manic patients in comparison to the other diagnostic groups, and additionally, that concentration was not high enough to produce this lethal effect. Taken together, all these factors meant that there was probably another unknown substance in urine that was responsible for the deaths of the animals. That unknown substance could either act independently or enhance the lethality effect of urea. His next step was to try combinations of the components in urine, but this led nowhere. In fact, he discovered that creatinine prevented convulsions and death caused by urea, and after further pursuing this pathway, Cade was able to show some anticonvulsant efficacy of creatinine [27]. It is not known whether this is valid since nobody ever tried to replicate these results. Another combination would be urea plus uric acid, but here, there was a practical problem: uric acid does not readily dissolve in water. One of the best solutions to this problem was to use the most soluble salt of uric acid, lithium urate. Again, surprisingly, lithium urate reduced urea toxicity, and when Cade decided to clarify whether this was due to lithium rather than to uric acid, he discovered that this was the case with other lithium salts like lithium carbonate, which also seemed to have protected guinea pigs from urea's toxicity. Further testing confirmed the protective effect of lithium salts. Cade also noted a suppressive and calming effect of lithium on animals and described the animals after lithium injection as 'extremely lethargic', but returning to normal after one or two hours. In fact, it is possible that the animals were lethargic because of lithium toxicity. All the facts suggest that these guinea pigs endured not a beneficial tranquilizing effect of lithium but rather early signs of lithium poisoning, which typically include lethargy and lack of responsiveness. Other investigators who later gave guinea pigs and other animals lithium were unable to reproduce the inactivity that Cade observed unless they injected the animals with large toxic doses.

The discovery of lithium as a potential therapeutic agent led naturally to the next step, which was to use it on patients. It is not known whether this decision was also supported by knowledge of the previous history of lithium as a treatment option during the late 19th century. It is a fact, however, that Cade 'arbitrarily' selected the doses of lithium that he initially gave to manic patients based on the doses that had been used during the previous century to patients suffering from diseases presumably caused by uric acid. This is similar to the pathway followed by Hammond since the dosages reported by both him and Cade lead to lithium intoxication. Interestingly, Cade himself suggests that he just got lucky. In reminiscing about his discovery 20 years later, he said, 'The original therapeutic dose decided on fortuitously proved to be the optimum, that is 1200 mg of the citrate thrice daily or 600 mg of the carbonate'. But given the effectiveness of the dose he decided on, some have speculated that along with Cade's good luck, a bit of conscious deliberation may have been at play.

Cade took lithium himself to see if it would cause adverse effects and to establish the correct dosage before administering it to patients. He took single and repeated doses of lithium citrate and lithium carbonate starting in early February 1948 and continued to do so for two weeks. Cade did not record what doses he consumed but recounted that they were the doses he thought would be appropriate to give to patients. First, he tried lithium on himself to establish how high a safe dose could be. Then, Cade began treating ten people with acute mania.

His first patient, known only by his initials, WB, and his first name (William or Bill), had suffered symptoms of mania and depression for 30 years and had been confined to Bundoora Hospital for the past five years with unrelenting symptoms of mania. After his excellent response to lithium and discharge from the hospital, he was readmitted six months later with severe mania. His brother reported that he had become overconfident about having been well for so many months, became reluctant to take medication, and eventually ceased taking lithium about six weeks before. Subsequently, he had become increasingly more irritable and erratic. Once lithium was readministered, within two weeks, he had again returned to normal, and he was released again a month later.

Over the following year, Cade went on to treat nine more manic patients with lithium and saw equally good results. In September 1949, he published a paper reporting both fast and dramatic improvements in all of them [28]. The most important observation was that while the majority of them had been in and out of Bundoora for years after the treatment with lithium, five of them had improved enough to return to their homes and families. Cade was 37 years old when his 1949 paper appeared. He never did any further research on lithium. The last survivor of the original 10 manic patients died of natural causes in 1980 at the age of 76. He had been on lithium for more than 30 years. John Cade died the same year.

Sadly, two years after the publication on the successful use of lithium, the first death because of lithium toxicity was reported in a Bipolar patient who otherwise responded extremely well to treatment. This first death was Cade's first patient, WB, who had a rocky time over the two years after the publication. He remained outside the hospital for months at a time and worked intermittently, but he did not take lithium regularly and vacillated between periods of mania when he was off lithium and stomach upset when he was on it. In March 1950, noting that WB still had 'dyspepsia, anorexia, and vomiting', Cade stopped his lithium. 'Under all circumstances', Cade wrote, 'it seems that he would be better off as a free restless case of mania rather than the dyspeptic, frail little man he looks on adequate lithium'. Then, a couple of weeks later, he noted, 'he was rapidly reverting to his manic phase and forgetting his dyspepsia'. One month later, in the face of WB's unrelenting mania, Cade seemed to change his mind. In his note of 3 May 1950, he wrote that WB 'has continued manic, restless, euphoric, noisy, dirty, mischievous, destructive, flight of ideas and thoroughly pleased with himself. His state seems as much a menace to life as any possible toxic effects of lithium. Therefore, recommence lithium citrate gr. 20 tds. Today'. Within two weeks, WB was 'quieter but miserable and asthenic'. Cade discontinued the lithium, but WB started to go downhill. Over the next few days, he became semicomatose and had three seizures. WB died on 23 May 1950. Cade recorded as the primary cause of death 'toxaemia due to lithium salts, therapeutically administered'. It is to be noted that he never properly reported this death and the probable cause. WB's name exists in the coroner's records of that year, and the cause of death recorded is lithium poisoning [29–31]. Cade's worry about toxicity prompted him to stop studying and using lithium soon after he completed his clinical trial and to discourage others from using it as well, so in 1950, he abandoned his experiments with lithium. Cade's 1949 paper went largely unnoticed at the time. Searching for an element with similar therapeutic properties but less toxic, Cade experimented with salts of rubidium, cerium, and strontium. None proved therapeutic.

Interestingly, WB is the only patient that Cade portrays in his accounts; as a result, WB has achieved a sort of prominence as a poster child for lithium, or at least Cade's poster child, despite his death. Schioldann wrote that Cade's presentation of the case of WB was eventually misleading since he failed to explicitly report his death because of lithium toxicity [32]. We will never know if Cade's silence on this matter was a deliberate attempt to conceal a painful truth or, in part or whole, an unintentional lapse.

While Cade was publishing his findings, Edward Trautner at the Department of Physiology and Pharmacology at the University of Melbourne, realized that he could measure blood levels of lithium using flame photometry. Charles Noack (1917–1969), a new psychiatrist on the staff of Melbourne's Mont Park Hospital and Trautner's former student, had heard of Cade's lithium work and was interested in examining what seemed like a promising treatment. Noack turned to Trautner for help in planning and implementing a lithium investigation. Trautner quickly understood the importance of the matter and that the management of toxicity would be the key. Together with Noack, they proceeded to carry out a seminal study, which is, however, not as much historically appreciated as it deserves, although it is one of the founding stones of lithium treatment. They started with the same lithium doses used by Cade and carefully monitored patients for signs of lithium toxicity. They reduced the dose or temporarily stopped the lithium entirely if symptoms of toxicity occurred. Mindful of Talbott's findings [33] on the relationship between lithium poisoning and serum levels, they decided, along with meticulous observation of patients' symptoms, to measure lithium serum levels. This was the first attempt to measure serum lithium levels during lithium treatment of mania. It was made possible thanks to Victor Wynn (1920–2006), one of the pioneers of the study of metabolism [34]. At that time, Wynn was a young Research Fellow in the Physiology Department. At that time, the available techniques to measure electrolytes were cumbersome, inaccurate, and time-consuming, so Wynn personally raised the money to buy a flame spectrophotometer, a new device recently introduced by the Beckman Instrument Company, which made it possible to measure electrolytes speedily and with high precision. This flame spectrophotometer, which Wynn and his group were using to quantify sodium and potassium in biological fluids, could also be used to measure lithium, a similar chemical. Trautner and Noac were thus able to measure lithium levels and carefully monitor and adjust the lithium dose [35]. Their study established several key points, all of which have withstood the test of time. First, lithium, just as Cade had said, alleviated symptoms of mania. None of the 100 patients in their study (more than 30 of whom had acute mania) suffered serious intoxication, and none died. All but one of their more than thirty manic patients improved, and many continued on lithium and remained well during more than a year of observation. Noack and Trautner also recorded lithium's preventive efficacy against mania, and in accord with the observation by Cade, they confirmed that it had little impact on depression or the fundamental symptoms of schizophrenia. Finally, Noack and Trautner showed that serum lithium levels could be practically monitored. The levels they observed were consistent with Talbott's, the average was about 1.0 mEq per liter, and the highest level they saw was 2.1 mEq per liter in a patient with no signs of poisoning. In this and subsequent studies, Trautner and his colleagues identified a specific lithium blood-level range necessary for both safety and effectiveness [36–38]. Although Trautner's work made it possible for lithium to be widely studied and prescribed, his contributions have barely been acknowledged.

While Cade decided to abandon lithium as a treatment option, Noack and Trautner's paper received the attention of the Danish physician Mogens Schou (1918–2005; Figure 9). Mogens Schou was not a psychiatrist but a physician who specialized in clinical chemistry and had observed a dramatic therapeutic effect of long-term lithium treatment in his younger brother. Erik Stromgren (1909–1993), head at that time of the Aarhus University psychiatric clinic in Risskov, asked him to undertake a randomized controlled trial of lithium in mania [39–41].

Figure 9. Mogens Schou (1918–2005) with permission from the INHN.

Schou proceeded to conduct the first controlled study of lithium in acute mania. He randomized acutely manic patients to lithium or placebo with a flip of a coin, and every two weeks, placebo and lithium alternated. Some of the patients entered the study using proper randomized controlled trial (RCT) rules. In 1954, he published the results, which made a significant impact [42]. The RCT is now the gold standard for treatment research; since the late twentieth century, it has been the most widely used method, and it is absolutely required to assess and confirm the value of new treatments. But in the early 1950s, when Schou did his lithium study, the RCT was a fairly new approach that had only recently been applied in medical research (the first published RCT, a study of streptomycin in tuberculosis, came out in 1948). Before this time, new treatments were assessed, if they were assessed at all, more or less the way Cade had evaluated lithium. Schou and his colleagues confirmed the value of lithium in treating mania by carrying out one of the first RCTs, if not the very first RCT, in psychiatry. The result, which verified lithium's unique healing properties, was a turning point for lithium and psychiatry [42]. Schou measured serum lithium levels, and he found that they usually ranged from 0.5 to 2.0 mEq per liter and were consistent with Noack and Trautner's observations in their first study. Again, in agreement with them, Schou did not observe a strong correlation between serum lithium and signs of toxicity [42]. Of utmost importance was his observation that the effect of lithium was not simple sedation but a true therapeutic effect on manic symptoms. Schou also recorded that some of the patients on long-term lithium treatment 'did not feel quite their own self', or, as one put it, felt 'kept down' by the medicine. Echoing that, some of today's patients complain that lithium blunts their feelings, perceptions, and creativity. However, lithium was still difficult to administer, and blood levels were a matter of guesswork. The situation changed with the introduction of the Coleman flame photometer in 1958, which made the monitoring of plasma lithium levels much more precise than the previously used Beckman photometer.

As for lithium's reception in Australia, by the mid-1950s, two more Australian doctors, in addition to Noack and Trautner, had published their experiences with lithium treatment. Bernard Glesinger, a psychiatrist at Claremont Hospital in Western Australia, gave lithium to 104 patients and found that 'the calming effect on maniacal, excited, hyperactive and restless patients was most satisfying and appreciable' [43] Another Australian doctor who had begun to use lithium, Max Margulies, of Lachlan Park Hospital in Tasmania, reported that several years earlier he had started treating 'maniacal' patients with lithium [44].

In a 1959 article that appeared in the journal Psychopharmacologia, Schou provided an update on lithium treatment with a paper under the title 'Lithium in Psychiatric Therapy: Stock-Taking after Ten Years' [45]. He reviewed the world's psychiatric literature for accounts of manic patients treated with lithium and, as of early 1959, found 15 published reports, including Cade's and his own. The total number of treated patients worldwide at that time was only 370. According to the authors of these reports, 304 of the patients (82%) had 'improved'. Schou acknowledged that this rate of improvement needed to be viewed cautiously; the authors varied in how they defined both mania and its improvement and for how long they treated and observed the patients. Except for that in Schou's study, the lithium treatment was 'uncontrolled', so it was not at all clear that lithium per se brought on the improvement.

The next year, Schou received letters from Geoffrey (Toby) Hartigan (1917–1968) from the UK and Poul Christian Baastrup (1918–2002; Figure 10) from Denmark. It is now known that for 25 years, Schou's younger brother had suffered yearly episodes of depression that lasted several months each, rendered him unable to work, and, despite hospital stays and treatment with Electroconvulsive therapy (ECT) and medication, would recur every spring. Schou wondered if lithium was worth trying. Hartigan was encouraging, Schou followed his suggestion, and the results were almost miraculous. As the years went by, a number of family members, in addition to his brother, received lithium treatment with good results. Schou found it enormously gratifying that his research with lithium, in addition to improving the lives of millions and shedding light on the nature of manic-depressive illness, was also of direct benefit to his family.

Figure 10. Left: Poul Christian Baastrup (1918–2002). Center: John Cade (1912–1980). Right: Mogens Schou (1918–2005). Copyright status: CC BY-SA 4.0.

A paper with Hartigan appeared in 1963 and was the first report to explicitly state that lithium could prevent severe depressive episodes [46,47]. Schou and Baastrup conducted a series of lithium experiments with ever stricter research rules. One year after Hartigan, in 1964, Baastrup published a report describing his first 11 patients and demonstrated the efficacy of lithium for the maintenance phase. In it, he mentioned Hartigan's related observations [48]. Determined to find out if lithium really prevents episodes of mania and depression, in 1960, Baastrup embarked on a prospective study. The paper appeared in the February 1967 issue of the *Archives of General Psychiatry* (today named *JAMA Psychiatry*). In it, Baastrup and Schou showed that lithium had a striking prophylactic effect. Before lithium treatment, relapses had occurred on average every 8 months. Lithium treatment reduced the rate of relapse to only every 60–85 months. Before treatment with lithium, patients were spending, on average, 13 weeks per year in a manic state. Lithium treatment reduced this to less than 2 weeks per year. Almost 80% of the patients had no relapses at all during lithium treatment [49]. Eventually, they conducted a double-blind, placebo-controlled clinical trial with a proper design. It was published in 1970 in *The Lancet*, and it established lithium as an effective treatment for most people with Bipolar disorder. The story of that paper is interesting. Schou and his collaborators sent the paper to *The Lancet*, but its editor rejected it, feeling that 'lithium had received enough publicity'. However, the intervention by W. Linford Rees, a prominent British psychiatrist, made the journal then agree to publish it, and it appeared in the 15 August 1970 issue [50].

In the meantime, in the US in 1960, Samuel Gershon received a Pfizer Fellowship that permitted him to join the Schizophrenia and Psycho-pharmacology Joint Research Project of the University of Michigan at the mental hospital in Ypsilanti, Michigan. While there, he carried his knowledge of lithium's benefits to his Michigan associates and treated about 20 patients with lithium. He also gave lectures about lithium at the National Institute of Mental Health and elsewhere. The same year, along with Arthur Yuwiler (1927–2012) [51], also at Ypsilanti, they published the first North American paper on lithium [52]. In 1965, Gershon became director of the Neuropsychopharmacology Research Unit at New York University. There, he established a lithium clinic and research programs devoted to lithium and Bipolar disorder. Gershon and his collaborators did a series of studies that supported lithium's specificity for the treatment of Bipolar disorder, identified some of lithium's important side effects, including interference with the activity of the thyroid gland, and compared lithium with other psychiatric drugs [53–55].

In 1958, Ronald Fieve (1930–2018) was in the midst of his psychiatric training at New York's Columbia University. He went on to carry out systematic studies of lithium as a treatment for both manic and depressive episodes. These studies confirmed lithium's value in the treatment of mania and also verified its lack of effectiveness as a remedy for acute depression [56–61].

Throughout the 1970s, Fieve took it upon himself to increase the public's awareness of manic-depressive illness and its successful treatment with lithium. He appeared on countless TV talk shows extolling the benefits of lithium, and in 1975, he published *Moodswing*, a book about manic depression for the lay public. It became a bestseller and sold over a million copies. In 1973, he invited Joshua Logan (1908–1988), the celebrated playwright, director, and producer (and also his patient), to appear with him on a televised American Medical Association (AMA) symposium about depression. Logan spoke candidly and in detail about his struggles with mania and depression and his recuperation with lithium. He went on to become a spokesperson for the value of lithium, and he and Fieve formed a persuasive duo on the talk show circuit. When, in 1973, they appeared with Barbara Walters on NBC's Today, the station was flooded with phone calls for weeks afterward. Logan even wrote a book about his personal experience of living with mental illness [62].

For the next few years, there was significant academic opposition to the use of lithium as the standard treatment for Bipolar disorder (BD), and much emphasis was given to its toxicity. Two Maudsley psychiatrists, Aubrey Lewis (1900–1975; Figure 11) and Michael Shepherd (1923–1995), claimed that the idea, based on Schou's studies, that lithium has a prophylactic effect was a 'myth'.

Figure 11. Aubrey Lewis (1900–1975), c. 1969, with permission from the National Portrait Gallery, London, https://npgimages.com/ (accessed on 7 April 2021).

They reproached Schou for not using a sufficiently rigorous control methodology. Some of the methodological fine points they raised were legitimate matters for debate and would be of particular relevance today. Aubrey Lewis was a professor of psychiatry and head of the Maudsley. He considered lithium treatment to be 'dangerous nonsense' while Michael Shepherd was also extremely negative towards it and suggested that lithium was toxic in mania and that claims of efficacy in the prevention of depression were based on 'dubious scientific methodology' [63–68]. The hard truth is that although later research confirmed the efficacy of lithium, at that time, in methodological terms, the Maudsley group was probably right. But the criticism took on an ad hominem tone. Critics also claimed that Schou was biased in favor of lithium's preventive benefit because he had treated his brother with lithium, who had suffered from intractable recurrent depression, and was convinced that lithium had cured him, transforming his life. As Schou recalled, he has been called a naïve and biased 'believer' [32]. Rumors circulated in the Maudsley group that Schou himself had manic depression and was on lithium. His critics hinted that this explained his enthusiasm for the drug, his seemingly unreasonable bias in favor of it, and his conviction that it was of value in the absence of a proper, controlled clinical trial. When Schou heard about this sort of whispering, he unequivocally denied that he either had manic-depressive or was taking lithium (and there was no evidence whatsoever that he was) [69]. Still, the ad hominem arguments persisted, and they hurt.

However, it is interesting that in the frame of this debate in the 1970s, the exact opposite opinion was also advocated for; that is, there was no need for strict methodology and statistics, and the clinical impression was overwhelming and sufficient [70]. The ensuing storm, known among psychopharmacology insiders as the 'Battle of Britain', was resolved over the ensuing decade only after further studies by Schou and others confirmed the initial findings. Lithium treatment for Bipolar disorder was approved in 1961 in France, in 1966 in the UK, in 1967 in Germany, and in 1970 in Italy. A practical problem was that since there was nothing to patent and little profit to be expected, no drug firm clamored to produce it, and there seemed to be no one to apply for labeling to the US authorities. The problem seemed so impossible that the American College of Neuropsychopharmacology considered filing an application in its own name to the Food and Drug Agency (FDA) so that lithium could be legally prescribed. This eventually became unnecessary when, finally, three Pharma companies submitted applications to the FDA for approval of lithium. In April 1970, with guidance from the lithium task force, the FDA authorized lithium for the treatment of manic episodes. By then, 49 countries had already approved lithium. The US was the 50th. In 1974, the FDA extended its approval of lithium to include its use as a preventive agent for prophylactic treatment of manic-depressive illness [20]. Following the success of lithium treatment, valproate was introduced in 1966 [71] and later carbamazepine was introduced [72].

With its initial approval for use in Bipolar patients, research boomed, and data were accumulating. More studies established lithium and robustly linked it to the treatment of all phases of Bipolar disorder [20,41,49,50,73–79]. Later, Fred Goodwin (1936–2020) suggested it could also be useful in the treatment of depression as an add-on to antidepressants [80–85]. The recommended serum lithium levels were determined with certainty only in 1976 [86].

In a paper published in 1963, Schou proposed that lithium and some of the new antidepressants might be a novel class of psychiatric drugs, ones that not only relieved some of the symptoms of manic-depressive illness but treated all of its features, the disease itself. He named these drugs 'mood-normalizers' or 'normothymotics', from the Greek 'θυμικό' ('thymiko', meaning emotionality or feelings) and 'θυμός' ('thymos', meaning anger) [87]. The term 'mood-normalizer' was taken after the term 'mood stabilizer', which was used during the 1950s to refer to a combination of amphetamine plus a barbiturate to treat patients with neurotic instability but not patients with Bipolar disorder.

The acceptance of lithium was slow, and there were several reasons for this. The main issue was that since lithium is a natural substance, the Pharma industry was not interested in it because it could not be patented. As a result, there was no commercial interest, and no drug company promoted it. At the same time, the industry intensively marketed the psychiatric drugs that came on the scene during the 1950s, and the psychiatric community focused on these new agents. As for the place of lithium in the lay culture, it is interesting and important to mention its relationship with *Maude*, a popular TV sitcom, that devoted two episodes (on 26 January and 2 February 1976; 'Maude's Mood', season 4, episodes 17 and 18) to manic-depressive illness. Norman Milton Lear (1922–2023), *Maude*'s producer, flew Nathan Kline to Los Angeles for a week to consult on the show. However, despite following Kline's advice, the original plot included the psychiatrist explaining that this disorder is a 'physical' problem, a 'chemical imbalance in the blood', usually easy to control with lithium, and he proceeded to treat Maude with lithium successfully. In the episodes broadcast, lithium was never mentioned; instead, a vague reference to medication was made. At the end of the second episode, the psychiatrist Dr. Herbert Lester (portrayed by Tim O'Connor) says that Maude's mood has been stabilized, so 'now we can search for the real causes', probably a necessary compromise to keep the prevailing psychoanalytical environment at least partially satisfied.

The advance publicity for the episodes mentioned Maude's good response to lithium. Although the show's reference to treatment was realistic and judicious, in 1976, lithium was still on shaky ground, and many psychiatrists, covering the spectrum from analysts to psychopharmacologists, objected to lithium's positive portrayal. The ensuing controversy received a good deal of press coverage, with many professionals declaring that lithium was dangerous. In an article by Les Brown in the *New York Times*, Sam Gershon stated that 'I have no serious reservations about lithium when used properly, but manic-depression is not a common disease, and that's what it should be used for'. He and others worried that lithium's favorable depiction on *Maude* could encourage a massive abuse of lithium and might encourage its use in all sorts of medical conditions (https://www.nytimes.com/1976/01/22/archives/lithium-use-in-maude-medical-issue.html) (accessed on 15 August 2025).

The most recent episode, however, took place in the 2010s. The hugely popular contemporary espionage thriller *Homeland*, which had its television debut in 2011 and finished its seventh season in 2018, features Carrie Mathison, a brilliant CIA agent played by Claire Danes, who has manic depression and handles her job and personal life well but only as long as she stays on lithium. In *Homeland*, lithium is not only clearly mentioned but it is also one of the key supporting players. Carrie confesses to a fellow agent: 'It keeps me safe and sane'. The show's depiction of her manic attacks, the crushing depressions that follow them, and lithium's critical role in her life have been praised for their authenticity by both health professionals and similarly affected patients.

4. Discussion

Lithium was an unlikely hero in the history of psychiatric pharmacotherapy. The timing of its re-discovery, the way it was discovered, and its position in lay culture make its story almost extraordinary. The very idea that a single element that essentially does not exist in the human body could be therapeutic for a major medical disorder seems almost unbelievable. The fact that it appeared because it could facilitate the dissolving in the water of a pathological accumulation in the joints in vitro is also remarkable. Its history also points to the great importance, personal histories, sensitivities, and family problems of researchers, present throughout the history of medicine. This is a pattern that emerges again and again, like the case of the brother of Mogens Schou, and often explains how unbelievable leaps in logical thinking were made, which eventually were proven 'lucky guesses'.

From a historical and sociopolitical standpoint, lithium constitutes one of the most valuable parts of psychiatry and one of its prides. It was the first material intervention proven efficacious in the treatment of mental disorders [88]. It received no funding or any kind of support from commercial interests and relied solely on the courage and persistence of researchers who worked against adverse conditions and even against the prevailing attitude of the time. Its use persists even today despite the lack of commercial interest and organized support.

Funding: This research received no external funding.

Data Availability Statement: Not applicable.

Conflicts of Interest: In the past, Dr Fountoulakis has received grants and served as a consultant, advisor, or CME speaker or received support for participating in congresses by AstraZeneca, Bristol-Myers Squibb, Eli Lilly, Ferrer, Gedeon Richter, Janssen, Lundbeck, Otsuka, Pfizer, the Pfizer Foundation, Sanofi-Aventis, Servier, Shire, and others. Since the beginning of 2020, he has had no conflict of interest at all.

References

1. Boesgaard, A.M.; Steigman, G. Big Bang Nucleosynthesis: Theories and Observations. *Annu. Rev. Astron. Astrophys.* **1985**, *23*, 319–378. [CrossRef]
2. National Center for Biotechnology Information. Periodic Table of Elements. 2025. Available online: https://pubchem.ncbi.nlm.nih.gov/periodic-table/ (accessed on 15 August 2025).
3. USGS. *Mineral Commodity Summaries 2025–Lithium*; United States Geological Survey (USGS): Reston, VA, USA, 2025.
4. D'Andrada, J. Des Caractères Et Des Propriétés De Plusieurs Nouveaux Minérauxde Suède Et De Norwège, Avec Quelques Observations Chimiques Faites Sur Ces Substances. *J. Chim. Phys.* **1800**, *51*, 239–245.
5. Krebs, R.E. *The History and Use of Our Earth's Chemical Elements: A Reference Guide*, 2nd ed.; Greenwood Press: Westport, CT, USA, 2006.
6. Berzelius, J.J. Ein Neues Mineralisches Alkali Und Ein Neues Metall [a New Mineral Alkali and a New Metal]. *J. Chem. Phys.* **1817**, *21*, 44–48.
7. Bunsen, R. Darstellung Des Lithiums [Preparation of Lithium]. *Ann. Chem. Pharm.* **1855**, *94*, 107–111. [CrossRef]
8. Garrett, D.E. *Handbook of Lithium and Natural Calcium Chloride*; Elsevier: Amsterdam, The Netherlands, 2004.
9. Mitchell, S.C.; Waring, R.H. Alexander Ure Md, Frcs (1808-1866), and the Beginning of Drug Metabolism Studies. *J. Med. Biogr.* **2024**, *32*, 119–124. [CrossRef]
10. Ure, A. Observations and Researches Upon a New Solvent for Stone in the Bladder. *Prov. Med. J. Retrosp. Med. Sci.* **1843**, *6*, 431. [CrossRef] [PubMed]
11. Ure, A. Calculus in the Bladder, Treated by Litholysis, or Solution of the Stone by Injections of the Carbonate of Lithia, Conjoined with Lithotrity (under the Care of Mr. Ure). *Lancet* **1860**, *1930*, 185–186. [CrossRef]
12. Symes, W.L. Haig on Headaches, Fits, and Mental Depression Due to the Abnormal Uric-Acid Elimination—T. W. Shore on the Morphology of the Vagus Nerve. *Brain* **1889**, *11*, 571–574. [CrossRef]
13. Whorton, J.C. Historical Development of Vegetarianism. *Am. J. Clin. Nutr.* **1994**, *59*, 1103S–1109S. [CrossRef]
14. Marmol, F. Lithium: Bipolar Disorder and Neurodegenerative Diseases Possible Cellular Mechanisms of the Therapeutic Effects of Lithium. *Prog. Neuro-Psychopharmacol. Biol. Psychiatry* **2008**, *32*, 1761–1771. [CrossRef]
15. Shorter, E. The History of Lithium Therapy. *Bipolar Disord.* **2009**, *11*, 4–9. [CrossRef]
16. Johnson, F.N.; Amdisen, A. The First Era of Lithium in Medicine. An Historical Note. *Pharmacopsychiatria* **1983**, *16*, 61–63. [CrossRef]
17. Strobusch, A.D.; Jefferson, J.W. The Checkered History of Lithium in Medicine. *Pharm. Hist.* **1980**, *22*, 72–76.
18. Mitchell, S.W. On the Use of Bromide of Lithium. *Am. J. Med. Sci.* **1870**, *60*, 443–445.
19. Clinical Lecture on Nervousness in the Male. *Med. News* **1877**, *35*, 177–184.
20. Mitchell, P.B.; Hadzi-Pavlovic, D. Lithium Treatment for Bipolar Disorder. *Bull. World Health Organ.* **2000**, *78*, 515.
21. Hammond, W.A. *A Treatise on Diseases of the Nervous System*; D. Appleton: New York, NY, USA, 1871.

22. Yeragani, V.K.; Gershon, S. Hammond and Lithium: Historical Update. *Biol. Psychiatry* **1986**, *21*, 1101–1102. [CrossRef]
23. Lenox, R.H.; Watson, D.G. Lithium and the Brain: A Psychopharmacological Strategy to a Molecular Basis for Manic Depressive Illness. *Clin. Chem.* **1994**, *40*, 309–314. [CrossRef] [PubMed]
24. Cade, J.F. John Frederick Joseph Cade: Family Memories on the Occasion of the 50th Anniversary of His Discovery of the Use of Lithium in Mania. 1949. *Aust. N. Z. J. Psychiatry* **1999**, *33*, 615–618. [CrossRef] [PubMed]
25. Burnet, F.M.; Cade, J.F.J.; Lush, D. The Serological, Response to Influenza Virus Infection During an Epidemic, with Particular Reference to Subclinical Infection. *Med. J. Aust.* **1940**, *1*, 397–401. [CrossRef]
26. Wallace, W.; de Moore, G. Edward Trautner (1890–1978), a Pioneer of Psychopharmacology. *J. Hist. Neurosci.* **2024**, *33*, 1–56. [CrossRef]
27. Cade, J.F. The Anticonvulsant Properties of Creatinine. *Med. J. Aust.* **1947**, *2*, 621–623. [CrossRef]
28. Cade, J.F. Lithium Salts in the Treatment of Psychotic Excitement. *Med. J. Aust.* **1949**, *2*, 349–352. [CrossRef] [PubMed]
29. Davies, B. The First Patient to Receive Lithium. 1983. *Aust. N. Z. J. Psychiatry* **1999**, *33*, 366–368. [CrossRef]
30. Davies, B. The First Patient to Receive Lithium. *Aust. N. Z. J. Psychiatry* **1983**, *17*, 366–368. [CrossRef]
31. Johnson, F.N.; John, F.J. Cade, 1912 to 1980: A Reminiscence. *Pharmacopsychiatria* **1981**, *14*, 148–149. [CrossRef]
32. Schioldann, N.J. *History of the Introduction of Lithium into Medicine and Psychiatry: Birth of Modern Psychopharmacology 1949*; Adelaide Academic Press [in collaboration with Brascoe Pub.]: Adelaide, Australia, 2009.
33. Talbott, J.H. Use of Lithium Salts as a Substitute for Sodium Chloride. *Arch. Intern. Med.* **1950**, *85*, 1–10. [CrossRef] [PubMed]
34. Richmond, C. Victor Wynn. *BMJ* **2007**, *334*, 431. [CrossRef]
35. Noack, C.H.; Trautner, E.M. The Lithium Treatment of Maniacal Psychosis. *Med. J. Aust.* **1951**, *2*, 219–222. [CrossRef]
36. Coats, D.A.; Trautner, E.M.; Gershon, S. The Treatment of Lithium Poisoning. *Australas. Ann. Med.* **1957**, *6*, 11–15. [CrossRef]
37. Trautner, E.M.; Morris, R.; Noack, C.H.; Gershon, S. The Excretion and Retention of Ingested Lithium and Its Effect on the Ionic Balance of Man. *Med. J. Aust.* **1955**, *42*, 280–291. [CrossRef]
38. Trautner, E.M.; Pennycuik, P.R.; Morris, R.J.; Gershon, S.; Shankly, K.H. The Effects of Prolonged Sub-Toxic Lithium Ingestion on Pregnancy in Rats. *Aust. J. Exp. Biol. Med. Sci.* **1958**, *36*, 305–321. [CrossRef]
39. Cole, N.; Parker, G. Cade's Identification of Lithium for Manic-Depressive Illness--the Prospector Who Found a Gold Nugget. *J. Nerv. Ment. Dis.* **2012**, *200*, 1101–1104. [CrossRef]
40. Mitchell, P.B.; Hadzi-Pavlovic, D. John Cade and the Discovery of Lithium Treatment for Manic Depressive Illness. *Med. J. Aust.* **1999**, *171*, 262–264. [CrossRef]
41. Bech, P. The Full Story of Lithium. A Tribute to Mogens Schou (1918–2005). *Psychother. Psychosom.* **2006**, *75*, 265–269. [CrossRef]
42. Schou, M.; Juel-Nielsen, N.; Stromgren, E.; Voldby, H. The Treatment of Manic Psychoses by the Administration of Lithium Salts. *J. Neurol. Neurosurg. Psychiatry* **1954**, *17*, 250–260. [CrossRef]
43. Glesinger, B. Evaluation of Lithium in Treatment of Psychotic Excitement. *Med. J. Aust.* **1954**, *41*, 277–283. [CrossRef] [PubMed]
44. Margulies, M. Suggestions for the Treatment of Schizophrenic and Manic-Depressive Patients. *Med. J. Aust.* **1955**, *42*, 137–141. [CrossRef] [PubMed]
45. Schou, M. Lithium in Psychiatric Therapy. Stock-Taking after Ten Years. *Psychopharmacologia* **1959**, *1*, 65–78. [CrossRef] [PubMed]
46. Jonhson, F.N. *The History of Lithium Therapy*; Oxford University Press: Oxford, UK, 1984.
47. Hartigan, G.P. The Use of Lithium Salts in Affective Disorders. *Br. J. Psychiatry* **1963**, *109*, 810–814. [CrossRef]
48. Baastrup, P.C. The Use of Lithium in Manic-Depressive Psychosis. *Compr. Psychiatry* **1964**, *5*, 396–408. [CrossRef] [PubMed]
49. Baastrup, P.C.; Schou, M. Lithium as a Prophylactic Agent–Its Effect against Recurrent Depressions and Manic-Depressive Psychosis. *Arch. Gen. Psychiatry* **1967**, *16*, 162–172. [CrossRef]
50. Baastrup, P.C.; Poulsen, J.C.; Schou, M.; Thomsen, K.; Amdisen, A. Prophylactic Lithium: Double Blind Discontinuation in Manic-Depressive and Recurrent-Depressive Disorders. *Lancet* **1970**, *2*, 326–330.
51. Gershon, S.; Oxenkrug, G. Arthur Yuwiler. *Neuropsychopharmacology* **2012**, *37*, 2887. [CrossRef]
52. Gershon, S.; Yuwiler, A. Lithium Ion: A Specific Psychopharmacological Approach to the Treatment of Mania. *J. Neuropsychiatry* **1960**, *1*, 229–241.
53. Gershon, S. Lithium in Mania. *Clin. Pharmacol. Ther.* **1970**, *11*, 168–187. [CrossRef]
54. Gershon, S.; Chengappa, K.N.; Malhi, G.S. Lithium Specificity in Bipolar Illness: A Classic Agent for the Classic Disorder. *Bipolar Disord.* **2009**, *11*, 34–44. [CrossRef]
55. Shopsin, B.; Gershon, S. *Lithium; Its Role in Psychiatric Research and Treatment, Edited by Samuel Gershon and Baron Shopsin*; Plenum Press: New York, NY, USA, 1973.
56. Fieve, R.R. Lithium Therapy at the Millennium: A Revolutionary Drug Used for 50 Years Faces Competing Options and Possible Demise. *Bipolar Disord.* **1999**, *1*, 67–70. [CrossRef]
57. Ban, T.A.; Healy, D.; Shorter, E. *The History of Psychopharmacology and the Cinp–as Told in Autobiography: The Triumph of Psychopharmacology and the Story of Cinp*; Collegium Internationale Neuro-Psychopharmaco: Zurich, Switzerland, 2010.

58. Fieve, R.R.; Platman, S.R.; Plutchik, R.R. The Use of Lithium in Affective Disorders. I. Acute Endogenous Depression. *Am. J. Psychiatry* **1968**, *125*, 487–491. [CrossRef]
59. Fieve, R.R.; Platman, S.H.; Plutchik, R.R. The Use of Lithium in Affective Disorders. Ii. Prophylaxis of Depression in Chronic Recurrent Affective Disorder. *Am. J. Psychiatry* **1968**, *125*, 492–498.
60. Fieve, R.R.; Platman, S. Lithium and Thyroid Function in Manic-Depressive Psychosis. *Am. J. Psychiatry* **1968**, *125*, 527–530. [CrossRef]
61. Fieve, R.R.; Platman, S.R. Prophylactic Lithium? *Lancet* **1968**, *2*, 830. [CrossRef]
62. Logan, J. *Josh: My Up and Down, In and Out Life*; Dell Publishing: New York, NY, USA, 1976.
63. Blackwell, B.; Shepherd, M. Prophylactic Lithium. Another Therapeutic Myth? An Examination of the Evidence to Date. *Lancet* **1968**, *1*, 968–971. [CrossRef]
64. Shepherd, M. A Prophylactic Myth. *Int. J. Psychiatry* **1970**, *9*, 423–425.
65. Blackwell, B. Need for Careful Evaluation of Lithium. *Am. J. Psychiatry* **1969**, *125*, 1131. [CrossRef]
66. Blackwell, B.; Schou, M. Lithium. *Lancet* **1970**, *2*, 875–876. [CrossRef]
67. Blackwell, B. Lithium Prophylaxis in Recurrent Affective Disorders. *Br. J. Psychiatry* **1971**, *1118*, 131–132. [CrossRef] [PubMed]
68. Blackwell, B. Prophylactic Lithium: Science or Science Fiction? *Am. Heart J.* **1972**, *83*, 139–141. [CrossRef] [PubMed]
69. Healy, D. *The Psychopharmacologists*; CRC Press: London, UK, 2018.
70. Sargant, W. Prophylactic Lithium? *Lancet* **1968**, *2*, 216. [CrossRef]
71. Lambert, P.A.; Carraz, G.; Borselli, S.; Carbel, S. [Neuropsychotropic Action of a New Anti-Epileptic Agent: Depamide]. *Ann. Med. Psychol.* **1966**, *124*, 707–710.
72. Okuma, T.; Inanaga, K.; Otsuki, S.; Sarai, K.; Takahashi, R.; Hazama, H.; Mori, A.; Watanabe, M. Comparison of the Antimanic Efficacy of Carbamazepine and Chlorpromazine: A Double-Blind Controlled Study. *Psychopharmacology* **1979**, *66*, 211–217. [CrossRef] [PubMed]
73. Schou, M.; Baastrup, P.C.; Grof, P.; Weis, P.; Angst, J. Pharmacological and Clinical Problems of Lithium Prophylaxis. *Br. J. Psychiatry* **1970**, *116*, 615–619. [CrossRef] [PubMed]
74. Angst, J.; Weis, P.; Grof, P.; Baastrup, P.C.; Schou, M. Lithium Prophylaxis in Recurrent Affective Disorders. *Br. J. Psychiatry* **1970**, *116*, 604–614. [CrossRef]
75. Angst, J.; Grof, P.; Schou, M. Lithium. *Lancet* **1969**, *1*, 1097. [CrossRef]
76. Schioldann, J. John Cade's Seminal Lithium Paper Turns Fifty. *Acta Psychiatr. Scand.* **1999**, *1*, 403–405.
77. Schioldann, J. 'On Periodical Depressions and Their Pathogenesis' by Carl Lange (1886). *Hist. Psychiatry* **2011**, *22*, 108–130. [CrossRef]
78. Schioldann, J. Obituary. *Hist. Psychiatry* **2006**, *17*, 247–252. [CrossRef]
79. Johnstone, E.C.; Crow, T.J.; Frith, C.D.; Owens, D.G. The Northwick Park Functional Psychosis Study: Diagnosis and Treatment Response. *Lancet* **1988**, *2*, 119–125. [CrossRef]
80. Goodwin, F.K.; Fireman, B.; Simon, G.E.; Hunkeler, E.M.; Lee, J.; Revicki, D. Suicide Risk in Bipolar Disorder During Treatment with Lithium and Divalproex. *JAMA* **2003**, *290*, 1467–1473. [CrossRef]
81. Goodwin, F.K. Rationale for Long-Term Treatment of Bipolar Disorder and Evidence for Long-Term Lithium Treatment. *J. Clin. Psychiatry* **2002**, *63*, 5–12.
82. Goodwin, F.K.; Zis, A.P. Lithium in the Treatment of Mania: Comparisons with Neuroleptics. *Arch. Gen. Psychiatry* **1979**, *36*, 840–844. [CrossRef] [PubMed]
83. Goodwin, F.K.; Murphy, D.L.; Dunner, D.L.; Bunney, W.E., Jr. Lithium Response in Unipolar Versus Bipolar Depression. *Am. J. Psychiatry* **1972**, *129*, 44–47. [CrossRef] [PubMed]
84. Goodwin, F.K.; Murphy, D.L.; Bunney, W.E., Jr. Lithium-Carbonate Treatment in Depression and Mania. A Longitudinal Double-Blind Study. *Arch. Gen. Psychiatry* **1969**, *21*, 486–496. [CrossRef]
85. Goodwin, F.K.; Murphy, D.L.; Bunney, W.E. Lithium. *Lancet* **1969**, *2*, 212–213. [CrossRef]
86. Bech, P.; Vendsborg, P.B.; Rafaelsen, O.J. Lithium Maintenance Treatment of Manic-Melancholic Patients: Its Role in the Daily Routine. *Acta Psychiatr. Scand.* **1976**, *53*, 70–81. [CrossRef]
87. Schou, M. Normothymotics, Mood-Normalizers: Are Lithium and the Imipramine Drugs Specific for Affective Disorders? *Br. J. Psychiatry* **1963**, *109*, 803–809. [CrossRef]
88. Fountoulakis, K.N.; Tohen, M.; Zarate, C.A. Lithium treatment of Bipolar disorder in adults: A systematic review of randomized trials and meta-analyses. *Eur. Neuropsychopharmacol.* **2022**, *54*, 100–115. [CrossRef] [PubMed]

pharmaceuticals

MDPI

Review

The Mechanisms of Lithium Action: The Old and New Findings

Kosma Sakrajda [1,*] and Janusz K. Rybakowski [2,*]

1 Molecular and Cell Biology Unit, Poznan University of Medical Sciences, 60-572 Poznan, Poland
2 Department of Adult Psychiatry, Poznan University of Medical Sciences, 60-572 Poznan, Poland
* Correspondence: ksakrajda@ump.edu.pl (K.S.); janusz.rybakowski@gmail.com (J.K.R.)

Abstract: Despite lithium's presence in modern psychiatry for three-quarters of a century, the mechanisms of its therapeutic action have not been fully elucidated. This article presents the evolution of the views on these mechanisms, and both the old and new findings are discussed. Among the old mechanisms, lithium's effect on the purinergic system; electrolyte metabolism; membrane transport; and second messenger systems, namely, cyclic nucleotide and phosphatidylinositol (PI), glycogen synthase kinase-3beta (GSK-3β), brain-derived neurotrophic factor, and neurotransmitters, are discussed. The new data were obtained from in vitro studies, molecular biology, and genetic research. They showed the effects of lithium on the immune system, biological rhythms, telomere functions, and mitochondria. In this article, each lithium mechanism is considered in the light of its association with the pathogenesis of bipolar disorder or/and as a marker of the lithium response. Although not exhaustive, this review elucidates the multiple potential mechanisms of lithium action. It was also observed that many seemingly "old" mechanisms have experienced a resurgence in research conducted during the 21st century. Additionally, many studies converged on the previously postulated mechanisms of lithium inhibiting GSK-3β and PI.

Keywords: lithium; bipolar disorder; lithium response

1. Introduction

In 2024, the 75th anniversary of introducing lithium into contemporary psychiatry was commemorated. This event was attested by an Australian psychiatrist John Cade's article in the 1949 issue of the *Medical Journal of Australia* [1]. Furthermore, in 1963, the first observations appeared pointing to lithium's property to prevent the recurrences of mood disorders [2]. Consequently, lithium is now the first-choice drug for preventing recurrences in mood disorders, mainly bipolar disorder (BD), as recommended in all major guidelines [3]. When treating an acute episode, lithium possesses strong antimanic and moderate antidepressant activity but is especially useful for the augmentation of antidepressant medications in treatment-resistant depression [4]. It was also demonstrated that lithium administration exerts suicide-preventive, immunomodulatory, and anti-dementia effects [5,6].

However, despite lithium's presence in modern psychiatry for three-quarters of a century, the exact mechanisms of its therapeutic action still have not been fully elucidated. In this paper, we present the evolution of perspectives on various selected mechanisms, discussing both the historical and recent findings and highlighting the pivotal roles played by several investigators in certain discoveries. Each mechanism is considered in light of its association with the pathogenesis of BD or/and as a marker of the lithium response.

2. Old Findings

2.1. Uric Acid and Purinergic System

2.1.1. Uric Acid

The influence of lithium treatment on uric acid levels, a crucial component of the purinergic system, can be regarded as the earliest proposed mechanism of its therapeutic action [7]. The first medical application of lithium was performed in 1859 by Alfred Garrod (1819–1907) to treat gout [8]. The rationale for using lithium was the previous observation of increased urate levels in patients with gout and the good solubility of the lithium urate. Therefore, lithium was meant to reduce the urates as lithium urate [8]. In the same context, in 1886, the Danish scientist Carl Lange (1834–1900) proposed uric acid diathesis as a pathogenic mechanism of periodical depression and suggested the therapeutic use of lithium carbonate as a treatment that allegedly normalizes the uric acid levels [9].

The concept of the pathogenic role of uric acid concerning mood disorders and lithium treatment also laid the foundation for the introduction of lithium to contemporary psychiatry by John Cade (1912–1980), who observed that the urine of manic patients was particularly toxic to guinea pigs. Cade attributed this to the excess of urates and administered lithium urate to the animals, obtaining a sedative effect, and subsequently also administered lithium carbonate with the same result. This effect further led him to treat manic patients with lithium carbonate with spectacular outcomes [1].

The observations of uric acid alterations in mood disorders might indicate the involvement of this substance in the pathogenesis of mood disorders and the effect of lithium. In 1968, it was shown that lithium had a uricosuric effect in patients with BD, and its excretion increased in the early phase of remission [10]. The 21st century brought further indications of uric acid's role in the pathogenesis of bipolar disorder. Taiwanese researchers showed that 16.4% of patients with bipolar disorder and 13.6% of the control group had gout. The risk of onset of gout during the 6-year observation period was 1.19 times greater for people with BD than for the control group [11]. Further clinical studies revealed that patients with a first episode of mania had increased concentrations of uric acid, which decreased as the clinical condition improved after one month of treatment [12,13]. However, the 2016 meta-analysis conducted by Bartoli et al. showed that patients with BD had significantly higher serum concentrations of uric acid as compared with healthy individuals and those suffering from periodic depression [14]. Interestingly, the study by Muti et al. showed that lithium-treated patients' serum lithium levels positively correlated with their uric acid levels [15]. Therefore, based on this result, the exact effect of lithium treatment on uric acid levels and the precise role of uric acid in BD should be further studied.

2.1.2. Purinergic System

Lithium might also influence the purinergic signaling system. The concept of this system was proposed in the 1970s by Geoffrey Burnstock (1929–2020), where the purine nucleotides adenosine 5′-triphosphate (ATP) and adenosine diphosphate (ADP) were defined as extracellular messengers [16].

Gubert et al. showed that in patients with BD, the adenosine concentration was lower compared with the control group, and they also demonstrated the negative correlation between the adenosine concentration and depression severity [17]. An experimental study also found that lithium increases the adenosine concentration by inhibiting ectonucleotidase [18].

The purinergic receptor P2X7 role was previously postulated in the pathogenesis of BD. This receptor mediates the processes of apoptosis and proliferation, mediates the release of proinflammatory cytokines, and is also involved in the mechanisms of neurotransmission and neuromodulation [19]. The P2X7 receptor gene is located on chromosome 12q23-24, which is described as a potential locus responsible for BD. An association between the gene polymorphism of that receptor and a predisposition to BD has also been described [20].

Lithium's effect on purinergic signaling can also be connected with the neuroprotective potential of the drug. Chronic lithium treatment inhibits ATP-induced cellular death by decreasing the ATP and increasing the adenosine levels [18]. The neuronal and microglial cells are responsive to in vitro ATP treatment, leading to ATP-mediated neural cell death and microglia phenotypes that respond to an inflammatory state. Lithium treatment results in the modulation of neuronal cell response by preventing cell death; however, it cannot prevent the microglia inflammatory response. This effect leads to the conclusion that the purinergic-based mechanism of lithium action is more likely to occur in neurons [21], but the exact purinergic-based neuroprotective effect should be further studied.

2.2. *Electrolyte Metabolism and Membrane Transport*

Lithium is a monovalent cation from the first group of the Mendeleev table, along with sodium and potassium. In the 1960s/1970s, interest was kindled in the possible effect of lithium on electrolyte metabolism and transmembrane transport in relation to its therapeutic mechanisms.

2.2.1. Lithium Accumulation in the Organism

An Australian psychiatrist, Maurice Serry, hypothesized that the therapeutic response to lithium can be proportional to the amount of retention of this ion in the organism. He elaborated on the lithium excretion test measuring the excretion of lithium within 4 h after loading a dose of 1200 mg of lithium carbonate [22]. He found that manic patients who were lithium retainers responded to lithium favorably, while the subjects that excessively excreted lithium did not [23]. It is considerable that the higher retention of lithium could be directly related to a decreased elimination of lithium from the cytosol, which is the topic of the following subsection.

2.2.2. Transmembrane Lithium Transport

In 1975, the primary mechanism of transporting lithium out of the cells by the lithium–sodium counter-transport system (LSC) was discovered. This discovery was made based on red blood cells by a team at Harvard Medical School led by Daniel Tosteson (1925–2009), the dean of the school from 1977 to 1997 [24]. Since the LSC determines the lithium concentration in some cells, e.g., in erythrocytes, its activity is related to the red blood cell lithium ratio. This index was further widely investigated concerning the pathogenesis of BD and therapeutic aspects of lithium treatment. A higher lithium blood cell index determines the ratio of lithium concentration in cells to that in serum. In patients with bipolar affective disorder, a higher concentration of lithium in the red blood cell was found in comparison with healthy subjects. This observation could indicate a reduced efflux of lithium from the blood cells governed by the LSC [25,26]. In other studies, the genetic component of the lithium ratio was established [27] and a lower rate of lithium transport from the cells in bipolar patients compared with control subjects was confirmed [28]. Unfortunately, the direct potential of the LSC mechanism for the pathogenesis of BD and the quality of lithium response has not been investigated in subsequent decades. However, the recent systematic review by Coyac et al. [29] of studies that researched the red blood cell lithium concentration and red blood cell/plasma lithium ratio in mood disorders concluded that further studies on these factors could be helpful as a novel method for monitoring lithium treatment tolerance.

2.2.3. Sodium–Potassium ATPase

In the 1970s/1980s, the findings of the decreased activity of the sodium–potassium-activated adenosine triphosphate pump (sodium–potassium ATPase) in erythrocytes of mood-disorder-suffering patients were reported, as well as the increased activity of this pump after lithium administration [30–33]. The sodium–potassium ATPase hypothesis for BD was formulated by El-Mallakh and Wyatt in 1995 [34]. This concept was further confirmed in later molecular genetic studies [35,36]. Therefore, the mechanism of lithium action in this context could be related to the pathogenesis of bipolar disorder. Unfortunately, studies on the effect of lithium on Na and K-ATPase have not been continued in recent decades.

2.3. Second Messenger Systems

Another mechanism potentially involved in lithium's modes of action is based on the second messenger systems. Second messenger systems modulate physiological processes by transmitting the signals from cell surface receptors (first messengers) to the intracellular effectors—enzymes, ion channels, and transporters. There are two major classes: cyclic nucleotide and phosphatidylinositol (PI) systems [37]. Both transpired to be connected with the pathogenesis of BD or/and lithium mechanisms.

2.3.1. Cyclic Nucleotide System

A synthesis of messenger cyclic monophosphate adenosine (cAMP) as a result of epinephrine-caused adenylyl cyclase stimulation was described by Earl Sutherland (1915–1974), a Nobel laureate in medicine in 1971 [38]. Further studies showed that G-protein subunits are signal transducers between receptors and adenylate cyclase [39]. The research team led by Robert Haim Belmaker showed that lithium treatment inhibits the adenyl cyclase activity and reduces the cAMP accumulation in the brain [40]. Chronic lithium treatment was also shown to interfere with the dissociation of the G-protein subunit Gi [41]. Furthermore, Harvey et al. [42] showed that chronic lithium treatment resulted in decreased cAMP levels caused by the increased activity of cAMP-phosphodiesterase and increased cGMP levels in rat's cortical regions.

In recent decades, evidence for the pathogenic role of adenylate cyclase in BD and the therapeutic mechanism of lithium was collected. In 2009, it was shown that lithium and carbamazepine preferentially inhibit specific isoforms—AC5, which might be connected with their antidepressant effect [43]. The genome-wide association study (GWAS) from 2014 revealed the adenylate cyclase-2 gene (*ADCY2*) as a risk gene for BD [44]. In 2020, Iranian researchers confirmed the association of the *ADCY2* polymorphism with BD and suggested it as a potential predictive marker of the lithium treatment response [45].

2.3.2. Phosphatidylinositol System

The PI system is crucial for receptor-mediated signal transduction. It involves the hydrolysis of phosphoinositides and the release of inositol trisphosphate (IP3) as a second messenger. The lithium effect on the PI system resulting in the uncompetitive inhibition of inositol monophosphatase (IMPase) was first demonstrated by American researchers from St Louis, Loretta Hallcher and William Sherman, in 1980 [46]. In 1989, the British physiologist and biochemist Michael Berridge (1938–2020) described the role of inositol trisphosphate (IP3) in cell signaling and calcium regulation [47]. He also further proved the uncompetitive inhibitory mechanism of lithium on inositol phosphate metabolism and formulated an inositol depletion hypothesis of BD. According to this hypothesis, lithium influences the PI system by weakening the second signaling. This proposed lithium mode of action includes the mentioned inhibition of the IMPase, which leads to increased inositol-1-monophosphate (I1P) levels, followed by phosphoinositols accumulation, decreased phosphatidylinositol concentration, and altered levels of key membrane phospholipids [47].

Lithium also lowers the phosphatidylinositol/phosphatidylcholine ratio in cell membranes [48]; decreases the amount of phosphatidylinositol-4,5-bisphosphate (PIP2) levels in the BD patient's platelet membrane; and reduces the phosphatidylinositol-specific phospholipase C (PIPLC), which is crucial for the generation of IP3, diacylglycerol, and also cAMP, resulting in the downregulation of inositol phosphates [49,50]. Israeli authors performed experimental studies on mice, where they deprived them of one of the two genes associated with the PI system: *IMPA1* (inositol monophosphatase-1) and *SMIT1* (sodium myo-inositol transporter-1). The study showed that both the genetic knock-out procedures caused behavior that resembled the effect after lithium administration, which may indicate that both inositol depletion and accumulation of phosphoinositols play an important role in the effects caused by lithium [51].

The 21st-century molecular–genetic studies suggest that the PI system may play a role in the pathogenesis of BD. A 2008 GWAS of BD patients identified an association with the di-acylglycerol kinase eTa (*DGKH*) gene, which encodes a crucial protein in the PI pathway [52]. In 2010, an association between polymorphism in the *IMPA2* gene and BD was found [53].

2.4. Glycogen Synthase Kinase-3 (GSK3) Activity

Glycogen synthase kinase-3 (GSK3) is a serine/threonine kinase that exists in two isoforms: GSK3 alpha (GSK3α) and GSK3 beta (GSK3β). It is involved in numerous signaling pathways, with over a hundred known substrates and multiple physiological functions. The GSK3β is connected, among others, with neurogenesis, neuronal polarization, axon growth, and biological rhythms [54].

The GSK3β is one of the best described molecular targets of lithium. In 1996, two independent studies by Peter Klein and Douglas Melton from the University of Pennsylvania [55] and Vuk Stambolic et al. from the University of Toronto [56] presented lithium's direct competitive mechanism of GSK3β inhibition. Also, the indirect method of lithium-caused inhibition was described as involving the phosphorylation of N-terminal serines of both GSK3α and GSK3β.

In a recent review, the four most important substrates of GSK3β were indicated as potentially contributing to the lithium effects in BD. They include the transcription factor cAMP response element-binding protein (CREB), the RNA-binding protein FXR1 (Fragile X related protein 1), kinesin subunits, and the cytoskeletal regulator CRMP2 (Collapsin response mediator protein 2). All these substrates are, in different ways, associated with the pathogenesis of BD [57]. Lithium's inhibition of GSK3β also influences the PI3K/Akt pathway, which leads to increased Akt-1 activity by inhibiting phosphoinositide 3-kinase-mediated Akt phosphorylation [58].

The next subsections show how many mechanisms of lithium can eventually be related to the inhibition of GSK3β. However, given the multiple substrates of this enzyme, it is hard to say which is most important to the drug's therapeutic effect in BD. In a recent GWAS that involved 40,000 BD subjects, no common variants of the *GSK3β* gene were associated with an increased risk for BD [59], whereas it was found that the lithium effect on GSK3β might play a role in lithium-induced renal decline [60] or in the anti-suicidal properties of this ion [61].

2.5. Brain-Derived Neurotrophic Factor

The brain-derived neurotrophic factor (BDNF) was identified in 1982 and quickly became the most explored neurotrophin in studies on the pathogenesis and treatment of mental disorders [62]. BDNF is essential for the function and survival of neurons and the regulation of neurotransmitters, such as glutamate, gamma-aminobutyric acid, dopamine, and serotonin. The BDNF system is important for the pathogenesis and treatment of mood disorders [63]. The Val66Met polymorphism of the *BDNF* gene is associated with a predilection to the BD illness [64].

In experimental studies, lithium enhanced the BDNF expression in the brain. One of the mechanisms behind this might be the stimulation of CREB, which further activates the *BDNF* gene expression [65]. Clinical studies showed that BDNF serum concentrations are reduced during episodes of both mania and depression and increase after successful pharmacological treatment, including lithium treatment. A low concentration of BDNF is also regarded as a marker of late-stage bipolar affective disorder [66]. However, our previous study from 2010 found that in the group of excellent lithium responders, despite the long duration of BD (20 years or more), the serum BDNF was not different from healthy subjects [67]. This supports the previous evidence regarding lithium's effect on BDNF expression and suggests that it restores the proper BDNF level. We also demonstrated that the Val allele of the BDNF Val66Met polymorphism in the BD patients' group determines better cognitive functions associated with the prefrontal cortex activity. This phenomenon was specific to bipolar disorder and was not observed in schizophrenic patients or healthy subjects [68]. Our previous study also showed the prophylactic potential of lithium treatment is connected with the Val66Met polymorphism of the *BDNF* [69].

2.6. Neurotransmitters

Since the 1960s, neurotransmitters, such as catecholamines (norepinephrine, dopamine) and serotonin, have been implicated in the pathogenesis of mood disorders and the therapeutic mechanism of antipsychotic and antidepressant drugs [70–72]. The effect of lithium on these neurotransmitters was also investigated.

2.6.1. Dopamine

In 2017, the latest variant of the dopamine hypothesis of BD was presented [73]. Hyperdopaminergia (elevations in D2/D3 receptor availability) was proposed to underline manic episodes, while increased dopamine transporter (DAT), resulting in reduced dopaminergic function, was suggested for depressive episodes. Since the 1970s, there has been evidence that lithium reduces dopaminergic activity, which may be connected with its antimanic effect [74]. Can et al. [75] demonstrated that chronic lithium treatment restored abnormal dopamine release in the nucleus accumbens. More recent reviews point to the regulation of dopamine transmission by lithium and the relationship of this effect to the lithium influence on GSK3β [76,77].

2.6.2. Serotonin

A hypothesis for the mechanism of long-term lithium action by stimulating serotonergic neurotransmission was proposed forty years ago [78]. The review of Price et al. [79] suggests that lithium's primary action on the serotonergic system may be presynaptic, with many postsynaptic effects leading to the enhancing effect on this neurotransmission. Recently, a study using positron emission tomography (PET) was performed with BD patients during a depressive episode. The patients with lower binding of the serotonin transporter and serotonin 5-HT1A receptor at the baseline showed more significant clinical improvement after the lithium treatment [80].

3. New Findings

3.1. In Vitro Studies

Research on lithium's mechanisms in recent years applied novel in vitro methods using peripheral blood mononuclear cells (PBMCs), lymphoblastoid cell lines (LCLs), and induced pluripotent stem cells (iPSCs). In most studies, the characteristics of cells obtained from so-called "lithium responders" (LRs) and "lithium non-responders" (LNRs) were compared.

3.1.1. Peripheral Blood Mononuclear Cells (PBMCs)

The study of human PBMCs from BD patients confirmed that lithium treatment results in the GSK-3β phosphorylation in Ser9, which can be evaluated as a biochemical marker of the therapeutic response to the drug [81]. It was also shown that PBMCs from BD patients and controls differ in immunophenotypes, showing immunologic imbalance in BD. The in vitro lithium treatment of PBMCs resulted in an increased percentage of CD14+ monocytes [82].

3.1.2. Lymphoblastoid Cell Lines (LCLs)

Lymphoblastoid cell lines (LCLs) are cell lines derived from B lymphocytes, typically immortalized using the Epstein–Barr virus (EBV) [83]. It was found that lithium did not alter the gene expression in the LCLs of healthy subjects but significantly changed the expression in the LCLs of BD patients—mostly the genes involved in apoptosis [84]. In an LCL-based genome-wide expression study, 2060 genes were differentially expressed between LR and LNR after in vitro lithium treatment. The further pathways analysis and qPCR validation showed insulin-like growth factor 1 as a significant biomarker of the lithium response [85]. Another study showed that the lithium treatment of LCLs from LR restored the cell viability and influenced the GSK-3β expression [86]. When comparing the RNA of LCLs from BD females, greater expressions of the *HDGFRP3* (hepatoma-derived growth factor-related protein-3) and *ID2* (inhibitor of DNA binding 2) genes were found in the LRs. These genes are connected with neurogenesis and neurotrophic processes [87]. The immunoglobulin-related genes in LCLs differed between BD patients and control subjects and between the LRs and LNRs [88].

Lithium in vitro treatment's effect on microRNA (miRNA) expression identified that seven miRNAs significantly changed after four days of treatment and four changed after 16 days of treatment, suggesting miRNAs as candidate's biomarker of lithium treatment [89]. Another study identified the differences in 335 mRNAs and 77 miRNAs between LRs and LNRs, mainly related to the immune processes, suggesting lithium's anti-inflammatory effects [90].

3.1.3. Pluripotent Stem Cells

Pluripotent stem cells (PSCs) are a type of cells characterized by the ability to self-renew and differentiate into any kind of cells found derived from the inner cell mass of embryos, primordial germ cells, teratocarcinoma, or male germ cells [91]. The studies of iPSC neurons derived from BD patients showed reversed hyperexcitability after the treatment only in neurons from LRs [92,93]. The cortical spheroids, or brain-like tissues derived from the iPSCs of BD patients, exhibited a smaller size and lower neuron excitability of the neurons compared with the healthy subjects. Treatment with lithium resulted in the normalization of the excitability of neurons to levels seen in controls [94]. The refurbishment of neuronal hyperactivity in LR was related to sodium currents and the Akt signaling pathway and to protein kinase B, which is connected with the growth and proliferation of the cell [95]. In LRs, neuronal progenitor cells derived from iPSCs demonstrated a high resistance to methamphetamine's toxic action [96].

On the other hand, in LNRs, a reduced expression of the *LEF1* gene (lymphoid enhancer-binding factor 1), which modulates neuronal activity, was shown [97]. Research on iPSC-derived cortical neurons corroborated the role of the PI system in lithium mechanisms. Inhibition by lithium of the IMPA1 enzyme, a crucial PI system element, was associated with reducing the neuronal excitability and releasing calcium ions [98].

3.2. Biological Rhythms

Circadian rhythms are intrinsic, 24 h cycles that regulates various physiological processes, including the sleep–wake cycle, body temperature regulation, and hormone secretion, and is driven by molecular clocks and synchronized with the external environment, primarily through light–dark cycles [99]. They are regulated by clock genes constituting feedback loops of transcription and translation. The positive feedback loop includes the aryl hydrocarbon receptor nuclear translocator-like (ARNTL) (also known as BMAL1) and circadian locomotor output cycles kaput (CLOCK). The CLOCK-BMAL1 complexes activate the expression of the nuclear receptors REV-ERBα and RORα and several clock genes, such as period (PER1/PER2) and cryptochrome (CRY1/CRY2) [100]. The animal model of bipolar mania is also based on *ClockΔ19* mice with a disrupted molecular circadian clock [101]. Interestingly, clinical research on BD patients revealed characteristics of biological rhythm disruptions, with a predominance of the eveningness chronotype [102]. The results from 95 studies showed that lithium delays the phase in various circadian rhythms (sleep–wakefulness cycles, activity rhythms, and peaks in elevation of body temperature) and also lengthens the circadian period [103]. The research on skin fibroblasts showed that LR patients more frequently exhibit the morningness chronotype [104] and that lithium treatment normalized circadian rhythm disturbances [105]. Corresponding results were shown in the study from research centers in Poznan and Cracow, Poland, where lithium-treated BD patients more often presented the morningness chronotype [106]. In the recent study by Mishra et al. [107], authors using a fibroblast-based model demonstrated a differential expression of circadian clock genes both between patients with BD and control persons, as well as between LRs and LNRs, with the meaningful distinction in the *Per 1* (clock protein PERIOD 1) and *Per 3* genes. Our study showed the connection with the effect of lithium prophylaxis was observed for six SNPs (single nucleotide polymorphisms) and three haplotype blocks of the *ARNTL* gene and two SNPs and one haplotype block of the *TIM* (timeless) gene [108].

The nuclear receptor REV-ERBα was also shown as a potential target for GSK3β and lithium [109]. In the mouse-based study, lithium-treated tissue slices and cells showed the elongation of period locomotor activity rhythms; lengthening of the molecular oscillations in the suprachiasmatic nucleus, lung tissue, and fibroblasts; and elevating *Per2* expression. The described effects might be caused by the lithium effect on GSK3β [110].

3.3. Telomere Functions

A protective cap protects the ends of chromosomes. This structure is called telomere (from the Greek words *telos*—end and *meros*—part), and its protective function is telomere capping. The 2009 Nobel Prize winners Elizabeth Blackburn, Carol Greider, and Jack Szostak explained the nature of telomeres and showed the mechanism behind chromosome protection by telomeres and the role of telomerase [111].

The telomere length can be regarded as a trait of aging processes. GWASs showed the genetic influence behind this effect, with the essential contribution of the leucine-rich repeat gene (*LRRC34*) found [112]. Shorter telomeres were also observed in patients with

BD, and their length was normalized after treatment with lithium [113–115]. It was also discovered that the duration of lithium treatment is positively associated with the leukocyte telomere length in BD patients [116] and positively correlated with the expression of the telomerase reverse transcriptase (TERT) enzyme [117]. On the other hand, the recent research, including 591 lithium-treated patients, did not reveal a relationship between the time of lithium administration and the length of telomeres [118].

Interestingly, the study on an animal model of depression showed that a lithium-naïve Flinders Sensitive Line rats presented with a shorter telomere length, downregulated telomerase reverse transcriptase expression, and reduced telomerase activity. Lithium treatment restored the telomerase activity and normalized the telomerase reverse transcriptase expression in the rat's hippocampus. This result might suggest a protective effect against telomere shortening and the neuroprotective role of lithium [119].

3.4. Immunomodulatory Effects

Granulopoiesis induction is a long-known effect of lithium treatment on the bone marrow due to increasing the granulocyte colony-stimulating factor [120,121]. Recent studies highlighted the immunomodulatory effect of lithium in the context of low-grade chronic inflammation and the microglial hypothesis of BD [122].

In a recent review by Sakrajda and Szczepankiewicz [123], evidence is presented for lithium's reduction of proinflammatory cytokines, i.e., Interleukin (IL)-1β, Interferon (IFN)-γ, Tumor Necrosis Factor (TNF), IL-6, and Monocyte Chemoattractant Protein (MCP)-1 expression, as well as the elevated expression of the anti-inflammatory cytokine IL-10. Lithium's action on the inflammatory system might be related to the lithium inhibition of GSK-3β. The GSK-3β activation is an important regulative mechanism of the NOD-like receptor protein 3 (NLRP3) inflammasome activation, which affects the production of inflammatory cytokines [124]. We previously showed that the mood-stabilizing treatment (with lithium or other drugs) of BD patients influences the inflammasome-regulated cytokines in euthymic BD patients. Still, only lithium-treated patients presented elevated expression of IL-10 [125]. Similar results were presented in the most recent review by Odeya Damri and Galila Agam [126], showing that chronic lithium treatment leads to altered cytokines expression in BD patients, decreases the secretion of inflammatory mediators in peripheral blood leukocytes, and also reduces the secretion of IL-1β in monocytes in vitro. Interestingly, the study by Sakrajda et al. [127], which used the in vitro model of neuroinflammation with the microglia cell line (HMC3) and focused on NLRP3 inflammasome involvement in BD, showed that lithium treatment with concentrations similar to those observed in the brains (0.5 mM) with microglia induced with innate-immune cytokines caused a significant anti-inflammatory effect mediated via GSK-3β inhibition, which affected NLRP3 inflammasome priming; its activation; and as a result, IL-1β protein secretion [127].

3.5. Antiviral Effects

The lithium immune-related mechanism also includes its antiviral actions. The first description was made by Julian Lieb in 1979, presenting two patients with recurrent herpes virus infections that went into remission during lithium treatment [128]. One year later, researchers from the University of Birmingham, using a hamster kidney model, demonstrated lithium's inhibition of the herpes simplex virus (HSV) [129].

A retrospective assessment of labial herpes recurrences in patients receiving prophylactic lithium was performed in joint research between the Department of Adult Psychiatry, Poznan University of Medical Sciences, and the Department of Psychiatry of the University of Pennsylvania. In the group of Polish patients, over the course of the lithium administration, the termination of herpes recurrences took place in 46% of patients with labial herpes and the overall reduction in recurrence incidence was 64%. The more pronounced effect was revealed in subjects with serum lithium concentrations > 0.65 mmol/L and red blood cell lithium concentrations >0.35 mmol/L. In the group of American lithium-treated BD patients, the recurrences of labial herpes, compared with the five-year pre-lithium period, were reduced by 73%. On the other hand, patients with recurrent depression that received antidepressant drugs showed no difference in the recurrences of herpes compared with the five-year pretreatment period [130].

The anti-herpes effect of lithium could be related to its prophylactic and therapeutic effect in dementia, as the infection with herpes simplex virus type 1 (HSV1) produces a significant risk factor for Alzheimer's disease [131].

The research on lithium and RNA viruses is concerned mainly with respiratory infections. In a retrospective analysis, a significant decrease in the average yearly rates of flu-like infections was found in patients that took lithium but not those that received antidepressants [130]. Also, a 28% decrease in respiratory infections, possibly partly due to the RNA viruses, was observed during long-term treatment with lithium [131].

Shortly after the outbreak of the COVID-19 epidemic, the antiviral potential of lithium on coronaviruses shown in experimental studies was reviewed [132]. An increasing hope was expressed that lithium could be useable in the anti-COVID-19 treatment [133]. Only one randomized clinical study found that lithium augmentation could reduce the severity of the COVID-19 course [134]. Moreover, the review of electronic records of 26,554 patients demonstrated that the frequency of infection was lower in subjects with therapeutic lithium levels (0.5–1.0 mmol/L) compared with those with lithium concentrations < 0.5 mmol/L [135].

3.6. Effect on Mitochondria

The upregulated production of oxidative stress markers leading to mitochondrial, dopamine system, and immune system dysfunction has been described in BD patients [136,137]. Lithium has been described as a modulator of the electron transport chains (ETCs) and oxidative phosphorylation (OXPHOS) pathway activity, enhancing oxidative phosphorylation in the frontal cortex of postmortem BD patients [138]. The in vitro study based on human neurons (SH-SY5Y) and chronic lithium treatment resulted in increased resistance to oxidative stress and enhanced glucose consumption and glycolysis activity [139]. Furthermore, a clinical study reported that lithium treatment increased the complex I activity and enhanced OXPHOS, positively correlating with plasma lithium levels in BD patients [140]. The 2018 study by Stacey et al. [141] presented the co-expression network analysis of RNA-seq data from lithium-treated BD patients, with an over-representation of genes involved in mitochondrial functioning (ETC and OXPHOS-related genes) and also the downregulation of these genes' expressions in LRs compared with poor lithium responders. Interestingly, the previously described lithium's inhibitory effect on the GSK-3β might also be involved in its influence on mitochondria, as this kinase was shown to interact, i.a., with mitochondrial proteins and respiratory chains and regulate the various mitochondrial functions [142].

3.7. Genetic Studies

In the 21st century, research emerged that explored the genetic foundations of lithium's therapeutic mechanisms and efficacy, where various methodologies were employed. In most studies, the lithium response was measured using the Retrospective Assessment of the Lithium Response Phenotype Scale, known as the Alda Scale [143].

3.7.1. Candidate Genes

In 2013, a review of candidate genes connected with lithium's prophylactic response appeared, written by one of the authors of this review [144]. Since then, some additions were also performed in the Poznan center that, e.g., showed the association of lithium efficacy with stress response genes [145]. In 2021, Senner et al. [146] found candidate genes that showed an association with lithium efficacy, where at least two studies listed six genes, namely, *GSK3β*, *BDNF*, serotonin transporter (*5HTTLPR*), calcium voltage-gated channel auxiliary subunit gamma 2 (*CACNG2*), nuclear receptor subfamily 1 group D member 1 (*NR1D1*), and dopamine receptor D1 (*DRD1*) genes.

3.7.2. Genome-Wide Association Studies (GWASs)

The first GWAS of the lithium response in patients with BD was performed by Perlis et al. in 2009 [147]. They assessed the five-year risk of recurrence in 1177 BD patients, including 458 treated with lithium. When comparing the LRs with healthy controls, a difference was found in the *GRIA2* gene, which codes the ionotropic AMPA glutamatergic receptor. In 2014, Chen et al. [148] investigated the population with Han Chinese ancestry and demonstrated the association of the lithium response with the gene encoding glutamate decarboxylase-like protein 1 (*GADL1*). Two years later, Song et al. [149], on a sample of 3874 lithium users from Sweden and the UK, showed an association between the lithium response and the *SEC14* and spectrin domains 1 (*SESTD1*) genes, which encode a synapse protein that directly binds to phospholipids. A GWAS was also performed, which recognized a region on chromosome 3, p21.1, with the gene *GNL3* (G protein nucleolar 3), as essential for the proliferation of neuronal progenitor cells by lithium [143].

3.7.3. ConLiGen Project

The Consortium of Lithium Genetics (ConLiGen) was established in 2009 as a collaborative initiative by the International Group for Study of Lithium-treated Patients (IGSLI) and the National Institute of Mental Health. The initial report about this event was published in 2010 [150]. The first results of a GWAS that concerned molecular–genetic factors connected with the preventive effect of lithium and involved 2563 patients from 22 sites were reported six years later. The association of the lithium effect was found with a region on chromosome 21, including two long non-coding RNAs (lncRNAs) that modulate the central nervous system's expression of the genes [151]. Subsequently, ConLiGen tried to assess the association between lithium efficacy in BD and the polygenic risk score (PRS) for different diseases. It turned out that the PRS characteristic for schizophrenia was linked to a poorer lithium effect [152], which correlated with research performed in Poznan, pointing to a negative connection between lithium's prophylactic effect and a predisposition to psychotic symptoms [153]. Negatively connected with lithium efficacy were also the PRS for major depressive disorder [154] and the PRS for attention deficit hyperactivity disorder (ADHD) [155]. A worse response to lithium was also associated with a higher PRS for diabetes and hypertension in bipolar 1 but not bipolar 2 patients [156].

The ConLiGen also assessed the relationship of lithium's prophylactic effect to human leukocyte antigen (HLA) genes. It turned out that a positive lithium effect was connected with HLA genes for decreased inflammation, while a negative effect was associated with HLA genes for increased inflammation [157]. Moreover, Amare et al. [152] recognized 36 candidate genes mainly connected with the cholinergic and glutamatergic systems, linked to a positive lithium effect. Recently, a collaborative study between the groups of ConLiGen and Pharmacogenomics of Bipolar Disorder demonstrated an association between lithium prophylaxis with focal adhesion and the PI3K-Akt signaling pathway, which are connected, i.a., with the proliferation and growth of the cell [158].

In 2023, ConLiGen's work based on a comprehensive analysis of genetic and clinical data from 2367 BD patients treated with lithium proposed the polygenic score for lithium treatment response (Li^{+}_{PGS}) [159]. The proposed biological marker showed statistically significant associations with the lithium treatment response, and Li^{+}_{PGS} might be a helpful pharmacogenomic tool to predict the patient's response to lithium therapy in the future.

3.7.4. Epigenetic Findings

The epigenetic studies of French researchers identified the biomarkers connected with the efficacy of lithium. The methylation differences between LRs and LNRs were disclosed using a genome-wide methylomic approach (SeqCapEpi) [160]. The combination of epigenetic factors with clinical ones, such as the first episode polarity, psychotic symptoms, and BD family load, made it possible to predict the effect of lithium for 86% of patients [161]. The confirmation of specific methylation patterns determining the lithium response in BD also came from a recent study that obtained 130 differentially methylated positions and 16 differentially methylated regions in LRs vs. LNRs [162].

4. Conclusions

Although certainly not complete, this review shows that the possible mechanisms of lithium action are multifold and still interesting research topics. In this review, the evolution of the views on these mechanisms starting in the mid-19th century is shown, while trying to distinguish between old and new findings. However, it turns out that many apparently "old" mechanisms had their revival in research performed in the 21st century. We also aimed to signal the potential links and overlaps between the proposed mechanisms of action, which might explain the widely described multimodality of lithium action. Notably, many eventually converged toward the lithium inhibition of GSK-3β and PI, the mechanisms postulated in the 1980s/1990s. In 2013, Brown and Tracy assumed these mechanisms to be the most important for lithium action at the cellular level [163], and in 2016, Malhi and Outhred simply stated that the research in the 21st century cemented GSK-3β inhibition as a key mechanism underpinning the effects of lithium [164]. In a recent review of the biological bases and treatment strategies in BD, the inhibition of the PI system and GSK-3β as the main mechanisms for lithium action are also underscored [165].

We also want to emphasize the high heterogeneity of the studies' results on different mechanisms mentioned in this review. In our opinion, this might have resulted from several factors, including the different study designs used, differences between the studied populations and between each patient, lack of longitudinal studies that observed the changes in the same subjects during chronic treatment, or differences in patient treatment.

While some of the lithium mechanisms discussed in our review are linked to the pathogenesis of BD, many of these relationships are, at best, tentative. On the other hand, certain findings associated with these mechanisms could serve as markers for the lithium response, and several are proposed as such. Advances in molecular genetics may enable the inclusion of genetic markers for this purpose, which was recently described in studies from ConLiGen, although the application of genetic markers in clinical practice may take time. However, based on lithium's multimodal action and the beneficial effect of good responsiveness to treatment, such integration would be invaluable for identifying the best candidates for long-term lithium therapy, which is the most effective method of BD prophylaxis, though it has recently been underutilized.

The main directions for repurposing lithium beyond BD are the augmentation of antidepressants in treatment-resistant depression [4] and the use in neurodegenerative disorders, such as Alzheimer's disease [166]. Interestingly, the effect of lithium augmentation of antidepressants was previously associated with a polymorphism of the *GSK-3β* gene [167]. In Alzheimer's disease, the mechanisms of lithium action may also involve the GSK-3β inhibition [168], lithium's effect on mitochondria [169], and its anti-herpes activity [129,131].

Therefore, it might be concluded that despite gathering substantial information on this subject, further studies are still necessary to understand lithium's modes of action.

Author Contributions: Conceptualization, J.K.R.; writing—original draft preparation, K.S. and J.K.R.; writing—review and editing, K.S. and J.K.R.; supervision, J.K.R. All authors have read and agreed to the published version of the manuscript.

Funding: This research received no external funding.

Conflicts of Interest: The authors declare no conflicts of interest.

References

1. Cade, J.F. Lithium Salts in the Treatment of Psychotic Excitement. *Med. J. Aust.* **1949**, *2*, 349–352. [CrossRef] [PubMed]
2. Hartigan, G.P. The Use of Lithium Salts in Affective Disorders. *Br. J. Psychiatry* **1963**, *109*, 810–814. [CrossRef] [PubMed]
3. Yatham, L.N.; Kennedy, S.H.; Parikh, S.V.; Schaffer, A.; Bond, D.J.; Frey, B.N.; Sharma, V.; Goldstein, B.I.; Rej, S.; Beaulieu, S.; et al. Canadian Network for Mood and Anxiety Treatments (CANMAT) and International Society for Bipolar Disorders (ISBD) 2018 Guidelines for the Management of Patients with Bipolar Disorder. *Bipolar Disord.* **2018**, *20*, 97–170. [CrossRef]
4. Bauer, M.; Adli, M.; Ricken, R.; Severus, E.; Pilhatsch, M. Role of Lithium Augmentation in the Management of Major Depressive Disorder. *CNS Drugs* **2014**, *28*, 331–342. [CrossRef] [PubMed]
5. Tondo, L.; Baldessarini, R.J. Prevention of Suicidal Behavior with Lithium Treatment in Patients with Recurrent Mood Disorders. *Int. J. Bipolar Disord.* **2024**, *12*, 6. [CrossRef]
6. Rybakowski, J.K. Antiviral, Immunomodulatory, and Neuroprotective Effect of Lithium. *J. Integr. Neurosci.* **2022**, *21*, 68. [CrossRef]
7. Mandal, A.K.; Mount, D.B. The Molecular Physiology of Uric Acid Homeostasis. *Annu. Rev. Physiol.* **2015**, *77*, 323–345. [CrossRef]
8. Garrod, A. *The Nature and Treatment of Gout, and Rheumatic Gout*; Walton & Maberly: London, UK, 1859.
9. Lange, C.G. *Om Periodiske Depressionstilstande Og Deres Patogenese: Foredrag Holdt i Medicinsk Selskab den 19. Januar 1886, Med en Efterskrift*; Jacob Lunds Forlag: Copenhagen, Denmark, 1886.
10. Anumonye, A.; Reading, H.W.; Knight, F.; Ashcroft, G.W. Uric-Acid Metabolism in Manic-Depressive Illness and during Lithium Therapy. *Lancet* **1968**, *291*, 1290–1293. [CrossRef] [PubMed]
11. Chung, K.H.; Huang, C.C.; Lin, H.C. Increased Risk of Gout among Patients with Bipolar Disorder: A Nationwide Population-Based Study. *Psychiatry Res.* **2010**, *180*, 147–150. [CrossRef] [PubMed]
12. Chatterjee, S.S.; Ghosal, S.; Mitra, S.; Mallik, N.; Ghosal, M.K. Serum Uric Acid Levels in First Episode Mania, Effect on Clinical Presentation and Treatment Response: Data from a Case Control Study. *Asian J. Psychiatry* **2018**, *35*, 15–17. [CrossRef] [PubMed]
13. Salvadore, G.; Viale, C.I.; Luckenbaugh, D.A.; Zanatto, V.C.; Portela, L.V.; Souza, D.O.; Zarate, C.A.; Machado-Vieira, R. Increased Uric Acid Levels in Drug-Naïve Subjects with Bipolar Disorder during a First Manic Episode. *Prog. Neuro-Psychopharmacol. Biol. Psychiatry* **2010**, *34*, 819–821. [CrossRef]
14. Bartoli, F.; Crocamo, C.; Mazza, M.G.; Clerici, M.; Carrà, G. Uric Acid Levels in Subjects with Bipolar Disorder: A Comparative Meta-Analysis. *J. Psychiatr. Res.* **2016**, *81*, 133–139. [CrossRef]
15. Muti, M.; Del Grande, C.; Musetti, L.; Marazziti, D.; Turri, M.; Cirronis, M.; Pergentini, I.; Corsi, M.; Dell'Osso, L.; Corsini, G.U. Serum Uric Acid Levels and Different Phases of Illness in Bipolar I Patients Treated with Lithium. *Psychiatry Res.* **2015**, *225*, 604–608. [CrossRef] [PubMed]
16. Burnstock, G. The Past, Present and Future of Purine Nucleotides as Signalling Molecules. *Neuropharmacology* **1997**, *36*, 1127–1139. [CrossRef] [PubMed]
17. Gubert, C.; Jacintho Moritz, C.E.; Vasconcelos-Moreno, M.P.; Quadros dos Santos, B.T.M.; Sartori, J.; Fijtman, A.; Kauer-Sant'Anna, M.; Kapczinski, F.; Battastini, A.M.O.; Magalhães, P.V.d.S. Peripheral Adenosine Levels in Euthymic Patients with Bipolar Disorder. *Psychiatry Res.* **2016**, *246*, 421–426. [CrossRef]
18. Wilot, L.C.; Bernardi, A.; Frozza, R.L.; Marques, A.L.; Cimarosti, H.; Salbego, C.; Rocha, E.; Battastini, A.M.O. Lithium and Valproate Protect Hippocampal Slices against ATP-Induced Cell Death. *Neurochem. Res.* **2007**, *32*, 1539–1546. [CrossRef]
19. Gubert, C.; Rodrigo Fries, G.; Wollenhaupt de Aguiar, B.; Ribeiro Rosa, A.; Busnello, V.; Ribeiro, L.; Bueno Morrone, F.; Maria Oliveira Battastini, A.; Kapczinski, F. The P2X7 Purinergic Receptor as a Molecular Target in Bipolar Disorder. *Neuropsychiatr. Neuropsychol.* **2013**, *8*, 1–7.

20. Gubert, C.; Andrejew, R.; Jacintho Moritz, C.E.; Dietrich, F.; Vasconcelos-Moreno, M.P.; dos Santos, B.T.M.Q.; Fijtman, A.; Kauer-Sant'Anna, M.; Kapczinski, F.; da Silva Magalhães, P.V.; et al. Bipolar Disorder and 1513A>C P2RX7 Polymorphism Frequency. *Neurosci. Lett.* **2019**, *694*, 143–147. [CrossRef] [PubMed]
21. Gubert, C.; Andrejew, R.; Figueiro, F.; Bergamin, L.; Kapczinski, F.; Magalhães, P.V.d.S.; Battastini, A.M.O. Lithium-Induced Neuroprotective Activity in Neuronal and Microglial Cells: A Purinergic Perspective. *Psychiatry Res.* **2021**, *295*, 113562. [CrossRef]
22. Serry, M. Lithium Retention and Response. *Lancet* **1969**, *1*, 1267–1268. [CrossRef] [PubMed]
23. Serry, M. The Lithium Excretion Test: I. Clinical Application and Interpretation. *Aust. N. Z. J. Psychiatry* **1969**, *3*, 390–394. [CrossRef]
24. Haas, M.; Schooler, J.; Tosteson, D.C. Coupling of Lithium to Sodium Transport in Human Red Cells. *Nature* **1975**, *258*, 425–427. [CrossRef]
25. Rybakowski, J.; Frazer, A.; Mendels, J.; Alan Ramsey, T. Prediction of the Lithium Ratio in Man by Means of an in Vitro Test. *Clin. Pharmacol. Ther.* **1977**, *22*, 465–469. [CrossRef]
26. Rybakowski, J.; Frazer, A.; Mendels, J.; Ramsey, T.A. Erythrocyte Accumulation of the Lithium Ion in Control Subjects and Patients with Primary Affective Disorder. *Commun. Psychopharmacol.* **1978**, *2*, 99–104. [PubMed]
27. Dorus, E.; Pandey, G.N.; Shaughnessy, R.; Davis, J.M. Lithium Transport across the RBC Membrane. A Study of Genetic Factors. *Arch. Gen. Psychiatry* **1980**, *37*, 80–81. [CrossRef] [PubMed]
28. Pandey, G.N.; Dorus, E.; Casper, R.C.; Janicek, P.; Davis, J.M. Lithium Transport in Red Cells of Patients with Affective Disorders. *Prog. Neuro-Psychopharmacol. Biol. Psychiatry* **1984**, *8*, 547–555. [CrossRef]
29. Coyac, M.; Jalabert, L.; Declèves, X.; Etain, B.; Bellivier, F. Relevance of Red Blood Cell Lithium Concentration in the Management of Lithium-Treated Bipolar and Unipolar Disorders: A Systematic Narrative Review. *Int. J. Bipolar Disord.* **2024**, *12*, 35. [CrossRef] [PubMed]
30. Hokin-Neaverson, M.; Jefferson, J.W. Deficient Erythrocyte NaK-ATPase Activity in Different Affective States in Bipolar Affective Disorder and Normalization by Lithium Therapy. *Neuropsychobiology* **1989**, *22*, 18–25. [CrossRef] [PubMed]
31. Rybakowski, J.; Potok, E.; Strzyzewski, W.; Chłopocka-Woźniak, M. The Effects of Lithium Treatment on Cation Transport Processes in Erythrocytes of Patients with Affective Illnesses. *Pol. J. Pharmacol. Pharm.* **1983**, *35*, 209–215.
32. Rybakowski, J.; Potok, E.; Strzyzewski, W. Decreased Activity of Ouabain-Dependent Sodium and Potassium Fluxes in Erythrocytes during Depression and Mania. *Act. Nerv. Super.* **1983**, *25*, 72–74.
33. Naylor, G.J.; Smith, A.H.W.; Dick, E.G.; Dick, D.A.T.; McHarg, A.M.; Chambers, C.A. Erythrocyte Membrane Cation Carrier in Manic-Depressive Psychosis. *Psychol. Med.* **1980**, *10*, 521–525. [CrossRef]
34. El-Mallakh, R.S.; Wyatt, R.J. The Na,K-ATPase Hypothesis for Bipolar Illness. *Biol. Psychiatry* **1995**, *37*, 235–244. [CrossRef]
35. Mynett-Johnson, L.; Murphy, V.; McCormack, J.; Shields, D.C.; Claffey, E.; Manley, P.; McKeon, P. Evidence for an Allelic Association between Bipolar Disorder and a Na+, K+ Adenosine Triphosphatase Alpha Subunit Gene (ATP1A3). *Biol. Psychiatry* **1998**, *44*, 47–51. [CrossRef]
36. Goldstein, I.; Lerer, E.; Laiba, E.; Mallet, J.; Mujaheed, M.; Laurent, C.; Rosen, H.; Ebstein, R.P.; Lichtstein, D. Association between Sodium- and Potassium-Activated Adenosine Triphosphatase Alpha Isoforms and Bipolar Disorders. *Biol. Psychiatry* **2009**, *65*, 985–991. [CrossRef] [PubMed]
37. Newton, A.C.; Bootman, M.D.; Scott, J. Second Messengers. *Cold Spring Harb. Perspect. Biol.* **2016**, *8*, a005926. [CrossRef] [PubMed]
38. Sutherland, E.W. Studies on the Mechanism of Hormone Action. *Science (1979)* **1972**, *177*, 401–408. [CrossRef]
39. Hepler, J.R.; Gilman, A.G. G Proteins. *Trends Biochem. Sci.* **1992**, *17*, 383–387. [CrossRef]
40. Newman, M.E.; Belmaker, R.H. Effects of Lithium in Vitro and Ex Vivo on Components of the Adenylate Cyclase System in Membranes from the Cerebral Cortex of the Rat. *Neuropharmacology* **1987**, *26*, 211–217. [CrossRef] [PubMed]
41. Colin, S.F.; Chang, H.C.; Mollner, S.; Pfeuffer, T.; Reed, R.R.; Duman, R.S.; Nestler, E.J. Chronic Lithium Regulates the Expression of Adenylate Cyclase and Gi-Protein Alpha Subunit in Rat Cerebral Cortex. *Proc. Natl. Acad. Sci. USA* **1991**, *88*, 10634–10637. [CrossRef] [PubMed]
42. Harvey, B.; Carstens, M.; Taljaard, J. Lithium Modulation of Cortical Cyclic Nucleotides: Evidence for the Yin-Yang Hypothesis. *Eur. J. Pharmacol.* **1990**, *175*, 129–136. [CrossRef] [PubMed]
43. Mann, L.; Heldman, E.; Bersudsky, Y.; Vatner, S.F.; Ishikawa, Y.; Almog, O.; Belmaker, R.H.; Agam, G. Inhibition of Specific Adenylyl Cyclase Isoforms by Lithium and Carbamazepine, but Not Valproate, May Be Related to Their Antidepressant Effect. *Bipolar Disord.* **2009**, *11*, 885–896. [CrossRef] [PubMed]
44. Mühleisen, T.W.; Leber, M.; Schulze, T.G.; Strohmaier, J.; Degenhardt, F.; Treutlein, J.; Mattheisen, M.; Forstner, A.J.; Schumacher, J.; Breuer, R.; et al. Genome-Wide Association Study Reveals Two New Risk Loci for Bipolar Disorder. *Nat. Commun.* **2014**, *5*, 3339. [CrossRef]
45. Aghabozorg Afjeh, S.S.; Shams, J.; Hamednia, S.; Boshehri, B.; Olfat, A.; Omrani, M.D. Investigation of the Impact of an ADCY2 Polymorphism as a Predictive Biomarker in Bipolar Disorder, Suicide Tendency and Response to Lithium Carbonate Therapy: The First Report from Iran. *Pharmacogenomics* **2020**, *21*, 1011–1020. [CrossRef] [PubMed]

46. Hallcher, L.M.; Sherman, W.R. The Effects of Lithium Ion and Other Agents on the Activity of Myo-Inositol-1-Phosphatase from Bovine Brain. *J. Biol. Chem.* **1980**, *255*, 10896–10901. [CrossRef] [PubMed]
47. Berridge, M.J.; Downes, C.P.; Hanley, M.R. Neural and Developmental Actions of Lithium: A Unifying Hypothesis. *Cell* **1989**, *59*, 411–419. [CrossRef]
48. Ding, D.; Greenberg, M.L. Lithium and Valproate Decrease the Membrane Phosphatidylinositol/Phosphatidylcholine Ratio. *Mol. Microbiol.* **2003**, *47*, 373–381. [CrossRef] [PubMed]
49. Soares, J.C.; Mallinger, A.G.; Dippold, C.S.; Frank, E.; Kupfer, D.J. Platelet Membrane Phospholipids in Euthymic Bipolar Disorder Patients: Are They Affected by Lithium Treatment? *Biol. Psychiatry* **1999**, *45*, 453–457. [CrossRef] [PubMed]
50. Soares, J.C.; Chen, G.; Dippold, C.S.; Wells, K.F.; Frank, E.; Kupfer, D.J.; Manji, H.K.; Mallinger, A.G. Concurrent Measures of Protein Kinase C and Phosphoinositides in Lithium-Treated Bipolar Patients and Healthy Individuals: A Preliminary Study. *Psychiatry Res.* **2000**, *95*, 109–118. [CrossRef] [PubMed]
51. Agam, G.; Bersudsky, Y.; Berry, G.T.; Moechars, D.; Lavi-Avnon, Y.; Belmaker, R.H. Knockout Mice in Understanding the Mechanism of Action of Lithium. *Biochem. Soc. Trans.* **2009**, *37*, 1121–1125. [CrossRef]
52. Baum, A.E.; Akula, N.; Cabanero, M.; Cardona, I.; Corona, W.; Klemens, B.; Schulze, T.G.; Cichon, S.; Rietschel, M.; Nöthen, M.M.; et al. A Genome-Wide Association Study Implicates Diacylglycerol Kinase Eta (DGKH) and Several Other Genes in the Etiology of Bipolar Disorder. *Mol. Psychiatry* **2008**, *13*, 197–207. [CrossRef]
53. Bloch, P.J.; Weller, A.E.; Doyle, G.A.; Ferraro, T.N.; Berrettini, W.H.; Hodge, R.; Lohoff, F.W. Association Analysis between Polymorphisms in the Myo-Inositol Monophosphatase 2 (IMPA2) Gene and Bipolar Disorder. *Prog. Neuro-Psychopharmacol. Biol. Psychiatry* **2010**, *34*, 1515–1519. [CrossRef] [PubMed]
54. Beurel, E.; Grieco, S.F.; Jope, R.S. Glycogen Synthase Kinase-3 (GSK3): Regulation, Actions, and Diseases. *Pharmacol. Ther.* **2015**, *148*, 114–131. [CrossRef]
55. Klein, P.S.; Melton, D.A. A Molecular Mechanism for the Effect of Lithium on Development. *Proc. Natl. Acad. Sci. USA* **1996**, *93*, 8455–8459. [CrossRef] [PubMed]
56. Stambolic, V.; Ruel, L.; Woodgett, J.R. Lithium Inhibits Glycogen Synthase Kinase-3 Activity and Mimics Wingless Signalling in Intact Cells. *Curr. Biol.* **1996**, *6*, 1664–1669. [CrossRef]
57. Chatterjee, D.; Beaulieu, J.M. Inhibition of Glycogen Synthase Kinase 3 by Lithium, a Mechanism in Search of Specificity. *Front. Mol. Neurosci.* **2022**, *15*, 1028963. [CrossRef] [PubMed]
58. Tian, N.; Kanno, T.; Jin, Y.; Nishizaki, T. Lithium Potentiates GSK-3β Activity by Inhibiting Phosphoinositide 3-Kinase-Mediated Akt Phosphorylation. *Biochem. Biophys. Res. Commun.* **2014**, *450*, 746–749. [CrossRef]
59. Mullins, N.; Forstner, A.J.; O'Connell, K.S.; Coombes, B.; Coleman, J.R.I.; Qiao, Z.; Als, T.D.; Bigdeli, T.B.; Børte, S.; Bryois, J.; et al. Genome-Wide Association Study of More than 40,000 Bipolar Disorder Cases Provides New Insights into the Underlying Biology. *Nat. Genet.* **2021**, *53*, 817–829. [CrossRef]
60. Rybakowski, J.K.; Abramowicz, M.; Szczepankiewicz, A.; Michalak, M.; Hauser, J.; Czekalski, S. The Association of Glycogen Synthase Kinase-3beta (GSK-3β) Gene Polymorphism with Kidney Function in Long-Term Lithium-Treated Bipolar Patients. *Int. J. Bipolar Disord.* **2013**, *1*, 8. [CrossRef]
61. Malhi, G.S.; Das, P.; Outhred, T.; Irwin, L.; Morris, G.; Hamilton, A.; Lynch, K.; Mannie, Z. Understanding Suicide: Focusing on Its Mechanisms through a Lithium Lens. *J. Affect. Disord.* **2018**, *241*, 338–347. [CrossRef] [PubMed]
62. Barde, Y.A.; Edgar, D.; Thoenen, H. Purification of a New Neurotrophic Factor from Mammalian Brain. *EMBO J.* **1982**, *1*, 549–553. [CrossRef]
63. Hashimoto, K. Brain-Derived Neurotrophic Factor as a Biomarker for Mood Disorders: An Historical Overview and Future Directions. *Psychiatry Clin. Neurosci.* **2010**, *64*, 341–357. [CrossRef] [PubMed]
64. Li, M.; Chang, H.; Xiao, X. BDNF Val66Met Polymorphism and Bipolar Disorder in European Populations: A Risk Association in Case-Control, Family-Based and GWAS Studies. *Neurosci. Biobehav. Rev.* **2016**, *68*, 218–233. [CrossRef]
65. Fukumoto, T.; Morinobu, S.; Okamoto, Y.; Kagaya, A.; Yamawaki, S. Chronic Lithium Treatment Increases the Expression of Brain-Derived Neurotrophic Factor in the Rat Brain. *Psychopharmacology* **2001**, *158*, 100–106. [CrossRef] [PubMed]
66. Kauer-Sant'Anna, M.; Kapczinski, F.; Andreazza, A.C.; Bond, D.J.; Lam, R.W.; Young, L.T.; Yatham, L.N. Brain-Derived Neurotrophic Factor and Inflammatory Markers in Patients with Early- vs. Late-Stage Bipolar Disorder. *Int. J. Neuropsychopharmacol.* **2009**, *12*, 447–458. [CrossRef] [PubMed]
67. Rybakowski, J.K.; Suwalska, A. Excellent Lithium Responders Have Normal Cognitive Functions and Plasma BDNF Levels. *Int. J. Neuropsychopharmacol.* **2010**, *13*, 617–622. [CrossRef] [PubMed]
68. Rybakowski, J.K.; Borkowska, A.; Skibinska, M.; Hauser, J. Illness-Specific Association of Val66met BDNF Polymorphism with Performance on Wisconsin Card Sorting Test in Bipolar Mood Disorder. *Mol. Psychiatry* **2006**, *11*, 122–124. [CrossRef] [PubMed]
69. Rybakowski, J.K.; Suwalska, A.; Skibinska, M.; Dmitrzak-Weglarz, M.; Leszczynska-Rodziewicz, A.; Hauser, J. Response to Lithium Prophylaxis: Interaction between Serotonin Transporter and BDNF Genes. *Am. J. Med. Genet. Part B Neuropsychiatr. Genet.* **2007**, *144*, 820–823. [CrossRef] [PubMed]

70. Schildkraut, J.J. The Catecholamine Hypothesis of Affective Disorders: A Review of Supporting Evidence. *Am. J. Psychiatry* **1965**, *122*, 509–522. [CrossRef] [PubMed]
71. Bunney, W.E. The Current Status of Research in the Catecholamine Theories of Affective Disorders. *Psychopharmacol. Commun.* **1975**, *1*, 599–609. [PubMed]
72. Coppen, A. Defects in Monoamine Metabolism and Their Possible Importance in the Pathogenesis of Depressive Syndromes. *Psychiatr. Neurol. Neurochir.* **1969**, *72*, 173–180.
73. Ashok, A.H.; Marques, T.R.; Jauhar, S.; Nour, M.M.; Goodwin, G.M.; Young, A.H.; Howes, O.D. The Dopamine Hypothesis of Bipolar Affective Disorder: The State of the Art and Implications for Treatment. *Mol. Psychiatry* **2017**, *22*, 666–679. [CrossRef] [PubMed]
74. Gallager, D.W.; Pert, A.; Bunney, W.E. Haloperidol-Induced Presynaptic Dopamine Supersensitivity Is Blocked by Chronic Lithium. *Nature* **1978**, *273*, 309–312. [CrossRef] [PubMed]
75. Can, A.; Frost, D.O.; Cachope, R.; Cheer, J.F.; Gould, T.D. Chronic Lithium Treatment Rectifies Maladaptive Dopamine Release in the Nucleus Accumbens. *J. Neurochem.* **2016**, *139*, 576–585. [CrossRef] [PubMed]
76. Beaulieu, J.M. Converging Evidence for Regulation of Dopamine Neurotransmission by Lithium: An Editorial Highlight for "Chronic Lithium Treatment Rectifies Maladaptive Dopamine Release in the Nucleus Accumbens". *J. Neurochem.* **2016**, *139*, 520–522. [CrossRef]
77. Mohamadian, M.; Fallah, H.; Ghofrani-Jahromi, Z.; Rahimi-Danesh, M.; Shokouhi Qare Saadlou, M.S.; Vaseghi, S. Mood and Behavior Regulation: Interaction of Lithium and Dopaminergic System. *Naunyn-Schmiedeberg's Arch. Pharmacol.* **2023**, *396*, 1339–1359. [CrossRef]
78. Muller-Oerlinghausen, B. Lithium Long-Term Treatment--Does It Act via Serotonin? *Pharmacopsychiatry* **1985**, *18*, 214–217. [CrossRef]
79. Price, L.H.; Charney, D.S.; Delgado, P.L.; Heninger, G.R. Lithium and Serotonin Function: Implications for the Serotonin Hypothesis of Depression. *Psychopharmacology* **1990**, *100*, 3–12. [CrossRef]
80. Ananth, M.; Bartlett, E.A.; DeLorenzo, C.; Lin, X.; Kunkel, L.; Vadhan, N.P.; Perlman, G.; Godstrey, M.; Holzmacher, D.; Ogden, R.T.; et al. Prediction of Lithium Treatment Response in Bipolar Depression Using 5-HTT and 5-HT1A PET. *Eur. J. Nucl. Med. Mol. Imaging* **2020**, *47*, 2417–2428. [CrossRef]
81. Li, X.; Friedman, A.B.; Zhu, W.; Wang, L.; Boswell, S.; May, R.S.; Davis, L.L.; Jope, R.S. Lithium Regulates Glycogen Synthase Kinase-3β in Human Peripheral Blood Mononuclear Cells: Implication in the Treatment of Bipolar Disorder. *Biol. Psychiatry* **2007**, *61*, 216–222. [CrossRef]
82. Wu, T.N.; Lee, C.S.; Wu, B.J.; Sun, H.J.; Chang, C.H.; Chen, C.Y.; Chen, C.K.; Wu, L.S.H.; Cheng, A.T.A. Immunophenotypes Associated with Bipolar Disorder and Lithium Treatment. *Sci. Rep.* **2019**, *9*, 17453. [CrossRef]
83. Ling, P.D.; Huls, H.M. Isolation and Immortalization of Lymphocytes. *Curr. Protoc. Mol. Biol.* **2005**, *70*, 28. [CrossRef]
84. Fries, G.R.; Colpo, G.D.; Monroy-Jaramillo, N.; Zhao, J.; Zhao, Z.; Arnold, J.G.; Bowden, C.L.; Walss-Bass, C. Distinct Lithium-Induced Gene Expression Effects in Lymphoblastoid Cell Lines from Patients with Bipolar Disorder. *Eur. Neuropsychopharmacol.* **2017**, *27*, 1110–1119. [CrossRef] [PubMed]
85. Squassina, A.; Costa, M.; Congiu, D.; Manchia, M.; Angius, A.; Deiana, V.; Ardau, R.; Chillotti, C.; Severino, G.; Calza, S.; et al. Insulin-like Growth Factor 1 (IGF-1) Expression Is Up-Regulated in Lymphoblastoid Cell Lines of Lithium Responsive Bipolar Disorder Patients. *Pharmacol. Res.* **2013**, *73*, 1–7. [CrossRef] [PubMed]
86. Paul, P.; Iyer, S.; Nadella, R.K.; Nayak, R.; Chellappa, A.S.; Ambardar, S.; Sud, R.; Sukumaran, S.K.; Purushottam, M.; Jain, S.; et al. Lithium Response in Bipolar Disorder Correlates with Improved Cell Viability of Patient Derived Cell Lines. *Sci. Rep.* **2020**, *10*, 7428. [CrossRef]
87. Milanesi, E.; Voinsky, I.; Hadar, A.; Srouji, A.; Maj, C.; Shekhtman, T.; Gershovits, M.; Gilad, S.; Chillotti, C.; Squassina, A.; et al. RNA Sequencing of Bipolar Disorder Lymphoblastoid Cell Lines Implicates the Neurotrophic Factor HRP-3 in Lithium's Clinical Efficacy. *World J. Biol. Psychiatry* **2019**, *20*, 449–461. [CrossRef] [PubMed]
88. Mizrahi, L.; Choudhary, A.; Ofer, P.; Goldberg, G.; Milanesi, E.; Kelsoe, J.R.; Gurwitz, D.; Alda, M.; Gage, F.H.; Stern, S. Immunoglobulin Genes Expressed in Lymphoblastoid Cell Lines Discern and Predict Lithium Response in Bipolar Disorder Patients. *Mol. Psychiatry* **2023**, *28*, 4280–4293. [CrossRef] [PubMed]
89. Chen, H.; Wang, N.; Burmeister, M.; McInnis, M.G. MicroRNA Expression Changes in Lymphoblastoid Cell Lines in Response to Lithium Treatment. *Int. J. Neuropsychopharmacol.* **2009**, *12*, 975–981. [CrossRef] [PubMed]
90. Cattane, N.; Courtin, C.; Mombelli, E.; Maj, C.; Mora, C.; Etain, B.; Bellivier, F.; Marie-Claire, C.; Cattaneo, A. Transcriptomics and MiRNomics Data Integration in Lymphoblastoid Cells Highlights the Key Role of Immune-Related Functions in Lithium Treatment Response in Bipolar Disorder. *BMC Psychiatry* **2022**, *22*, 665. [CrossRef]
91. Kuijk, E.W.; Chuva de Sousa Lopes, S.M.; Geijsen, N.; Macklon, N.; Roelen, B.A.J. The Different Shades of Mammalian Pluripotent Stem Cells. *Hum. Reprod. Update* **2011**, *17*, 254–271. [CrossRef]

92. Mertens, J.; Wang, Q.W.; Kim, Y.; Yu, D.X.; Pham, S.; Yang, B.; Zheng, Y.; Diffenderfer, K.E.; Zhang, J.; Soltani, S.; et al. Differential Responses to Lithium in Hyperexcitable Neurons from Patients with Bipolar Disorder. *Nature* **2015**, *527*, 95–99. [CrossRef] [PubMed]
93. Stern, S.; Santos, R.; Marchetto, M.C.; Mendes, A.P.D.; Rouleau, G.A.; Biesmans, S.; Wang, Q.W.; Yao, J.; Charnay, P.; Bang, A.G.; et al. Neurons Derived from Patients with Bipolar Disorder Divide into Intrinsically Different Sub-Populations of Neurons, Predicting the Patients' Responsiveness to Lithium. *Mol. Psychiatry* **2018**, *23*, 1453–1465. [CrossRef] [PubMed]
94. Osete, J.R.; Akkouh, I.A.; Ievglevskyi, O.; Vandenberghe, M.; de Assis, D.R.; Ueland, T.; Kondratskaya, E.; Holen, B.; Szabo, A.; Hughes, T.; et al. Transcriptional and Functional Effects of Lithium in Bipolar Disorder IPSC-Derived Cortical Spheroids. *Mol. Psychiatry* **2023**, *28*, 3033–3043. [CrossRef]
95. Khayachi, A.; Abuzgaya, M.; Liu, Y.; Jiao, C.; Dejgaard, K.; Schorova, L.; Kamesh, A.; He, Q.; Cousineau, Y.; Pietrantonio, A.; et al. Molecular Signatures of Hyperexcitability and Lithium Responsiveness in Bipolar Disorder Patient Neurons Provide Alternative Therapeutic Strategies. *bioRxiv* **2023**. [CrossRef]
96. Mishra, H.K.; Mandyam, A.D.; Trenet, W.; Wei, H.; Nievergelt, C.M.; Maihofer, A.X.; Shilling, P.D.; Alda, M.; Gershon, E.; McInnis, M.G.; et al. Neural Progenitor Cells Derived from Lithium Responsive and Non-Responsive Bipolar Disorder Patients Exhibit Distinct Sensitivity to Cell Death Following Methamphetamine. *Neuropharmacology* **2023**, *226*, 109410. [CrossRef]
97. Santos, R.; Linker, S.B.; Stern, S.; Mendes, A.P.D.; Shokhirev, M.N.; Erikson, G.; Randolph-Moore, L.; Racha, V.; Kim, Y.; Kelsoe, J.R.; et al. Deficient LEF1 Expression Is Associated with Lithium Resistance and Hyperexcitability in Neurons Derived from Bipolar Disorder Patients. *Mol. Psychiatry* **2021**, *26*, 2440–2456. [CrossRef] [PubMed]
98. Saha, S.; Krishnan, H.; Raghu, P. IMPA1 Dependent Regulation of Phosphatidylinositol 4,5-Bisphosphate and Calcium Signalling by Lithium. *Life Sci. Alliance* **2023**, *7*, e202302425. [CrossRef]
99. Grandin, L.D.; Alloy, L.B.; Abramson, L.Y. The Social Zeitgeber Theory, Circadian Rhythms, and Mood Disorders: Review and Evaluation. *Clin. Psychol. Rev.* **2006**, *26*, 679–694. [CrossRef] [PubMed]
100. Jagannath, A.; Taylor, L.; Wakaf, Z.; Vasudevan, S.R.; Foster, R.G. The Genetics of Circadian Rhythms, Sleep and Health. *Hum. Mol. Genet.* **2017**, *26*, 128–138. [CrossRef] [PubMed]
101. Roybal, K.; Theobold, D.; Graham, A.; DiNieri, J.A.; Russo, S.J.; Krishnan, V.; Chakravarty, S.; Peevey, J.; Oehrlein, N.; Birnbaum, S.; et al. Mania-like Behavior Induced by Disruption of CLOCK. *Proc. Natl. Acad. Sci. USA* **2007**, *104*, 6406–6411. [CrossRef] [PubMed]
102. Freund, N.; Haussleiter, I. Bipolar Chronobiology in Men and Mice: A Narrative Review. *Brain Sci.* **2023**, *13*, 738. [CrossRef] [PubMed]
103. Moreira, J.; Geoffroy, P.A. Lithium and Bipolar Disorder: Impacts from Molecular to Behavioural Circadian Rhythms. *Chronobiol. Int.* **2016**, *33*, 351–373. [CrossRef]
104. McCarthy, M.J.; Wei, H.; Nievergelt, C.M.; Stautland, A.; Maihofer, A.X.; Welsh, D.K.; Shilling, P.; Alda, M.; Alliey-Rodriguez, N.; Anand, A.; et al. Chronotype and Cellular Circadian Rhythms Predict the Clinical Response to Lithium Maintenance Treatment in Patients with Bipolar Disorder. *Neuropsychopharmacology* **2019**, *44*, 620–628. [CrossRef] [PubMed]
105. Federoff, M.; McCarthy, M.J.; Anand, A.; Berrettini, W.H.; Bertram, H.; Bhattacharjee, A.; Calkin, C.V.; Conroy, C.; Coryell, W.H.; D'Arcangelo, N.; et al. Correction of Depression-Associated Circadian Rhythm Abnormalities Is Associated with Lithium Response in Bipolar Disorder. *Bipolar Disord.* **2022**, *24*, 521–529. [CrossRef] [PubMed]
106. Dopierała, E.; Chrobak, A.; Tereszko, A.; Rybakowski, J. Lithium Influence on Circadian Rhythm Assessed by the Composite Scale of Morningness in Remitted Bipolar Patients. *Pharmakother. Psychiatry Neurol.* **2017**, *33*, 9–20.
107. Mishra, H.K.; Wei, H.; Rohr, K.E.; Ko, I.; Nievergelt, C.M.; Maihofer, A.X.; Shilling, P.D.; Alda, M.; Berrettini, W.H.; Brennand, K.J.; et al. Contributions of Circadian Clock Genes to Cell Survival in Fibroblast Models of Lithium-Responsive Bipolar Disorder. *Eur. Neuropsychopharmacol.* **2023**, *74*, 1–14. [CrossRef]
108. Rybakowski, J.K.; Dmitrzak-Weglarz, M.; Kliwicki, S.; Hauser, J. Polymorphism of Circadian Clock Genes and Prophylactic Lithium Response. *Bipolar Disord.* **2014**, *16*, 151–158. [CrossRef] [PubMed]
109. Yin, L.; Wang, J.; Klein, P.S.; Lazar, M.A. Nuclear Receptor Rev-Erbα Is a Critical Lithium-Sensitive Component of the Circadian Clock. *Science (1979)* **2006**, *311*, 1002–1005. [CrossRef]
110. Li, J.; Lu, W.Q.; Beesley, S.; Loudon, A.S.I.; Meng, Q.J. Lithium Impacts on the Amplitude and Period of the Molecular Circadian Clockwork. *PLoS ONE* **2012**, *7*, e33292. [CrossRef]
111. Blackburn, E.H.; Greider, C.W.; Szostak, J.W. Telomeres and Telomerase: The Path from Maize, Tetrahymena and Yeast to Human Cancer and Aging. *Nat. Med.* **2006**, *12*, 1133–1138. [CrossRef] [PubMed]
112. Coutts, F.; Palmos, A.B.; Duarte, R.R.R.; de Jong, S.; Lewis, C.M.; Dima, D.; Powell, T.R. The Polygenic Nature of Telomere Length and the Anti-Ageing Properties of Lithium. *Neuropsychopharmacology* **2018**, *44*, 757–765. [CrossRef] [PubMed]
113. Martinsson, L.; Wei, Y.; Xu, D.; Melas, P.A.; Mathe, A.A.; Schalling, M.; Lavebratt, C.; Backlund, L. Long-Term Lithium Treatment in Bipolar Disorder Is Associated with Longer Leukocyte Telomeres. *Transl. Psychiatry* **2013**, *3*, e261. [CrossRef] [PubMed]

114. Pisanu, C.; Congiu, D.; Manchia, M.; Caria, P.; Cocco, C.; Dettori, T.; Frau, D.V.; Manca, E.; Meloni, A.; Nieddu, M.; et al. Differences in Telomere Length between Patients with Bipolar Disorder and Controls Are Influenced by Lithium Treatment. *Pharmacogenomics* **2020**, *21*, 533–540. [CrossRef] [PubMed]
115. Ferensztajn-Rochowiak, E.; Kurczewska, E.; Rubiś, B.; Lulkiewicz, M.; Hołysz, H.; Rybakowski, F.; Rybakowski, J.K. Decreased Leucocyte Telomere Length in Male Patients with Chronic Bipolar Disorder: Lack of Effect of Long-Term Lithium Treatment. *Acta Neuropsychiatr* **2021**, *33*, 299–306. [CrossRef] [PubMed]
116. Squassina, A.; Pisanu, C.; Congiu, D.; Caria, P.; Frau, D.; Niola, P.; Melis, C.; Baggiani, G.; Lopez, J.P.; Cruceanu, C.; et al. Leukocyte Telomere Length Positively Correlates with Duration of Lithium Treatment in Bipolar Disorder Patients. *Eur. Neuropsychopharmacol.* **2016**, *26*, 1241–1247. [CrossRef]
117. Lundberg, M.; Millischer, V.; Backlund, L.; Martinsson, L.; Stenvinkel, P.; Sellgren, C.M.; Lavebratt, C.; Schalling, M. Lithium and the Interplay Between Telomeres and Mitochondria in Bipolar Disorder. *Front. Psychiatry* **2020**, *11*, 586083. [CrossRef] [PubMed]
118. Mutz, J.; Wong, W.L.E.; Powell, T.R.; Young, A.H.; Dawe, G.S.; Lewis, C.M. The Duration of Lithium Use and Biological Ageing: Telomere Length, Frailty, Metabolomic Age and All-Cause Mortality. *Geroscience* **2024**, *46*, 5981–5994. [CrossRef] [PubMed]
119. Wei, Y.B.; Backlund, L.; Wegener, G.; Mathé, A.A.; Lavebratt, C. Telomerase Dysregulation in the Hippocampus of a Rat Model of Depression: Normalization by Lithium. *Int. J. Neuropsychopharmacol.* **2015**, *18*, pyv002. [CrossRef] [PubMed]
120. Ferensztajn-Rochowiak, E.; Rybakowski, J.K. The Effect of Lithium on Hematopoietic, Mesenchymal and Neural Stem Cells. *Pharmacol. Rep.* **2016**, *68*, 224–230. [CrossRef]
121. Radomski, J.L.; Fuyat, H.N.; Nelson, A.A.; Smith, P.K. The toxic effects, excretion and distribution of lithium chloride. *J. Pharmacol. Exp. Ther.* **1950**, *100*, 429–444. [CrossRef] [PubMed]
122. Rosenblat, J.D.; McIntyre, R.S. Bipolar Disorder and Inflammation. *Psychiatr. Clin. N. Am.* **2016**, *39*, 125–137. [CrossRef]
123. Sakrajda, K.; Szczepankiewicz, A. Inflammation-Related Changes in Mood Disorders and the Immunomodulatory Role of Lithium. *Int. J. Mol. Sci.* **2021**, *22*, 1532. [CrossRef]
124. Arumugam, S.; Qin, Y.; Liang, Z.; Han, S.N.; Boodapati, S.L.T.; Li, J.; Lu, Q.; Flavell, R.A.; Mehal, W.Z.; Ouyang, X. GSK3β Mediates the Spatiotemporal Dynamics of NLRP3 Inflammasome Activation. *Cell Death Differ.* **2022**, *29*, 2060–2069. [CrossRef] [PubMed]
125. Sakrajda, K.; Bilska, K.; Czerski, P.M.; Narożna, B.; Dmitrzak-Węglarz, M.; Heilmann-Heimbach, S.; Brockschmidt, F.F.; Herms, S.; Nöthen, M.M.; Cichon, S.; et al. Abelson Helper Integration Site 1 Haplotypes and Peripheral Blood Expression Associates with Lithium Response and Immunomodulation in Bipolar Patients. *Psychopharmacology* **2024**, *241*, 727–738. [CrossRef]
126. Damri, O.; Agam, G. Lithium, Inflammation and Neuroinflammation with Emphasis on Bipolar Disorder—A Narrative Review. *Int. J. Mol. Sci.* **2024**, *25*, 13277. [CrossRef] [PubMed]
127. Sakrajda, K.; Langwiński, W.; Stachowiak, Z.; Ziarniak, K.; Narożna, B.; Szczepankiewicz, A. Immunomodulatory Effect of Lithium Treatment on in Vitro Model of Neuroinflammation. *Neuropharmacology* **2025**, *265*, 110238. [CrossRef] [PubMed]
128. Lieb, J. Remission of Recurrent Herpes Infection during Therapy with Lithium. *N. Engl. J. Med.* **1979**, *301*, 942. [CrossRef] [PubMed]
129. Skinner, G.R.B.; Hartley, C.; Buchan, A.; Harper, L.; Gallimore, P. The Effect of Lithium Chloride on the Replication of Herpes Simplex Virus. *Med. Microbiol. Immunol.* **1980**, *168*, 139–148. [CrossRef] [PubMed]
130. Rybakowski, J.K.; Amsterdam, J.D. Lithium Prophylaxis and Recurrent Labial Herpes Infections. *Lithium* **1991**, *2*, 43–47.
131. Itzhaki, R.F. Overwhelming Evidence for a Major Role for Herpes Simplex Virus Type 1 (HSV1) in Alzheimer's Disease (AD); Underwhelming Evidence Against. *Vaccines* **2021**, *9*, 679. [CrossRef] [PubMed]
132. Nowak, J.K.; Walkowiak, J.; Wei, F.; Wang, W.; Beaulieu, J.-M.; Chuang, D.-M. Lithium and Coronaviral Infections. A Scoping Review. *F1000Research* **2020**, *9*, 93. [CrossRef] [PubMed]
133. Murru, A.; Manchia, M.; Hajek, T.; Nielsen, R.E.; Rybakowski, J.K.; Sani, G.; Schulze, T.G.; Tondo, L.; Bauer, M. Lithium's Antiviral Effects: A Potential Drug for CoViD-19 Disease? *Int. J. Bipolar Disord.* **2020**, *8*, 21. [CrossRef] [PubMed]
134. Spuch, C.; López-García, M.; Rivera-Baltanás, T.; Cabrera-Alvargonzález, J.J.; Gadh, S.; Rodrigues-Amorim, D.; Álvarez-Estévez, T.; Mora, A.; Iglesias-Martínez-Almeida, M.; Freiría-Martínez, L.; et al. Efficacy and Safety of Lithium Treatment in SARS-CoV-2 Infected Patients. *Front. Pharmacol.* **2022**, *13*, 850583. [CrossRef]
135. De Picker, L.J.; Leboyer, M.; Geddes, J.R.; Morrens, M.; Harrison, P.J.; Taquet, M. Association between Serum Lithium Level and Incidence of COVID-19 Infection. *Br. J. Psychiatry* **2022**, *221*, 425–427. [CrossRef] [PubMed]
136. Madireddy, S.; Madireddy, S. Therapeutic Interventions to Mitigate Mitochondrial Dysfunction and Oxidative Stress–Induced Damage in Patients with Bipolar Disorder. *Int. J. Mol. Sci.* **2022**, *23*, 1844. [CrossRef]
137. Lam, X.J.; Xu, B.; Yeo, P.L.; Cheah, P.S.; Ling, K.H. Mitochondria Dysfunction and Bipolar Disorder: From Pathology to Therapy. *IBRO Neurosci. Rep.* **2023**, *14*, 407. [CrossRef] [PubMed]
138. Maurer, I.C.; Schippel, P.; Volz, H.P. Lithium-Induced Enhancement of Mitochondrial Oxidative Phosphorylation in Human Brain Tissue. *Bipolar Disord.* **2009**, *11*, 515–522. [CrossRef] [PubMed]

139. Nciri, R.; Desmoulin, F.; Allagui, M.S.; Murat, J.-C.; El Feki, A.; Vincent, C.; Croute, F. Neuroprotective Effects of Chronic Exposure of SH-SY5Y to Low Lithium Concentration Involve Glycolysis Stimulation, Extracellular Pyruvate Accumulation and Resistance to Oxidative Stress. *Int. J. Neuropsychopharmacol.* **2013**, *16*, 365–376. [CrossRef] [PubMed]
140. De Sousa, R.T.; Streck, E.L.; Zanetti, M.V.; Ferreira, G.K.; Diniz, B.S.; Brunoni, A.R.; Busatto, G.F.; Gattaz, W.F.; Machado-Vieira, R. Lithium Increases Leukocyte Mitochondrial Complex I Activity in Bipolar Disorder during Depressive Episodes. *Psychopharmacology* **2015**, *232*, 245–250. [CrossRef] [PubMed]
141. Stacey, D.; Schubert, K.O.; Clark, S.R.; Amare, A.T.; Milanesi, E.; Maj, C.; Leckband, S.G.; Shekhtman, T.; Kelsoe, J.R.; Gurwitz, D.; et al. A Gene Co-Expression Module Implicating the Mitochondrial Electron Transport Chain Is Associated with Long-Term Response to Lithium Treatment in Bipolar Affective Disorder. *Transl. Psychiatry* **2018**, *8*, 183. [CrossRef]
142. Yang, K.; Chen, Z.; Gao, J.; Shi, W.; Li, L.; Jiang, S.; Hu, H.; Liu, Z.; Xu, D.; Wu, L. The Key Roles of GSK-3β in Regulating Mitochondrial Activity. *Cell. Physiol. Biochem.* **2017**, *44*, 1445–1459. [CrossRef]
143. Grof, P.; Duffy, A.; Cavazzoni, P.; Grof, E.; Garnham, J.; MacDougall, M.; O'Donovan, C.; Alda, M. Is Response to Prophylactic Lithium a Familial Trait? *J. Clin. Psychiatry* **2002**, *63*, 942–947. [CrossRef]
144. Rybakowski, J.K. Genetic Influences on Response to Mood Stabilizers in Bipolar Disorder: Current Status of Knowledge. *CNS Drugs* **2013**, *27*, 165–173. [CrossRef]
145. Szczepankiewicz, A.; Narozna, B.; Rybakowski, J.K.; Kliwicki, S.; Czerski, P.; Dmitrzak-Węglarz, M.; Skibińska, M.; Twarowska-Hauser, J.; Pawlak, J. Genes Involved in Stress Response Influence Lithium Efficacy in Bipolar Patients. *Bipolar Disord.* **2018**, *20*, 753–760. [CrossRef]
146. Senner, F.; Kohshour, M.O.; Abdalla, S.; Papiol, S.; Schulze, T.G. The Genetics of Response to and Side Effects of Lithium Treatment in Bipolar Disorder: Future Research Perspectives. *Front. Pharmacol* **2021**, *12*, 638882. [CrossRef] [PubMed]
147. Perlis, R.H.; Smoller, J.W.; Ferreira, M.A.R.; McQuillin, A.; Bass, N.; Lawrence, J.; Sachs, G.S.; Nimgaonkar, V.; Scolnick, E.M.; Gurling, H.; et al. A Genomewide Association Study of Response to Lithium for Prevention of Recurrence in Bipolar Disorder. *Am. J. Psychiatry* **2009**, *166*, 718–725. [CrossRef] [PubMed]
148. Chen, C.-H.; Lee, C.-S.; Lee, M.-T.M.; Ouyang, W.-C.; Chen, C.-C.; Chong, M.-Y.; Wu, J.-Y.; Tan, H.K.-L.; Lee, Y.-C.; Chuo, L.-J.; et al. Variant GADL1 and Response to Lithium Therapy in Bipolar I Disorder. *N. Engl. J. Med.* **2014**, *370*, 119–128. [CrossRef] [PubMed]
149. Song, J.; Bergen, S.E.; Di Florio, A.; Karlsson, R.; Charney, A.; Ruderfer, D.M.; Stahl, E.A.; Chambert, K.D.; Moran, J.L.; Gordon-Smith, K.; et al. Genome-Wide Association Study Identifies SESTD1 as a Novel Risk Gene for Lithium-Responsive Bipolar Disorder. *Mol. Psychiatry* **2016**, *21*, 1290–1297. [CrossRef] [PubMed]
150. Schulze, T.G.; Alda, M.; Adli, M.; Akula, N.; Ardau, R.; Bui, E.T.; Chillotti, C.; Cichon, S.; Czerski, P.; Del Zompo, M.; et al. The International Consortium on Lithium Genetics (ConLiGen): An Initiative by the NIMH and IGSLI to Study the Genetic Basis of Response to Lithium Treatment. *Neuropsychobiology* **2010**, *62*, 72–78. [CrossRef] [PubMed]
151. Hou, L.; Heilbronner, U.; Degenhardt, F.; Adli, M.; Akiyama, K.; Akula, N.; Ardau, R.; Arias, B.; Backlund, L.; Banzato, C.E.M.; et al. Genetic Variants Associated with Response to Lithium Treatment in Bipolar Disorder: A Genome-Wide Association Study. *Lancet* **2016**, *387*, 1085–1093. [CrossRef] [PubMed]
152. Amare, A.T.; Schubert, K.O.; Hou, L.; Clark, S.R.; Papiol, S.; Heilbronner, U.; Degenhardt, F.; Tekola-Ayele, F.; Hsu, Y.H.; Shekhtman, T.; et al. Association of Polygenic Score for Schizophrenia and HLA Antigen and Inflammation Genes with Response to Lithium in Bipolar Affective Disorder: A Genome-Wide Association Study. *JAMA Psychiatry* **2018**, *75*, 65–74. [CrossRef] [PubMed]
153. Dembińska-Krajewska, D.; Kliwicki, S.; Chłopocka-Woźniak, M.; Rybakowski, J. The Effectiveness of Prophylactic Use of Lithium in Bipolar Disorder and Schizotypal Traits (in Polish). *Pharmakother Psychiatr Neurol.* **2012**, *28*, 153–158.
154. Amare, A.T.; Schubert, K.O.; Hou, L.; Clark, S.R.; Papiol, S.; Cearns, M.; Heilbronner, U.; Degenhardt, F.; Tekola-Ayele, F.; Hsu, Y.H.; et al. Association of Polygenic Score for Major Depression with Response to Lithium in Patients with Bipolar Disorder. *Mol Psychiatry* **2021**, *26*, 2457–2470. [CrossRef]
155. Coombes, B.J.; Millischer, V.; Batzler, A.; Larrabee, B.; Hou, L.; Papiol, S.; Heilbronner, U.; Adli, M.; Akiyama, K.; Akula, N.; et al. Association of Attention-Deficit/Hyperactivity Disorder and Depression Polygenic Scores with Lithium Response: A Consortium for Lithium Genetics Study. *Complex Psychiatry* **2021**, *7*, 80–89. [CrossRef]
156. Herrera-Rivero, M.; Adli, M.; Akiyama, K.; Akula, N.; Amare, A.T.; Ardau, R.; Arias, B.; Aubry, J.M.; Backlund, L.; Bellivier, F.; et al. Exploring the Genetics of Lithium Response in Bipolar Disorders. *Int. J. Bipolar Disord.* **2024**, *12*, 20. [CrossRef]
157. Le Clerc, S.; Lombardi, L.; Baune, B.T.; Amare, A.T.; Schubert, K.O.; Hou, L.; Clark, S.R.; Papiol, S.; Cearns, M.; Heilbronner, U.; et al. HLA-DRB1 and HLA-DQB1 Genetic Diversity Modulates Response to Lithium in Bipolar Affective Disorders. *Sci. Rep.* **2021**, *11*, 70. [CrossRef] [PubMed]
158. Ou, A.H.; Rosenthal, S.B.; Adli, M.; Akiyama, K.; Akula, N.; Alda, M.; Amare, A.T.; Ardau, R.; Arias, B.; Aubry, J.M.; et al. Lithium Response in Bipolar Disorder Is Associated with Focal Adhesion and PI3K-Akt Networks: A Multi-Omics Replication Study. *Transl. Psychiatry* **2024**, *14*, 109. [CrossRef]

159. Amare, A.T.; Thalamuthu, A.; Schubert, K.O.; Fullerton, J.M.; Ahmed, M.; Hartmann, S.; Papiol, S.; Heilbronner, U.; Degenhardt, F.; Tekola-Ayele, F.; et al. Association of Polygenic Score and the Involvement of Cholinergic and Glutamatergic Pathways with Lithium Treatment Response in Patients with Bipolar Disorder. *Mol. Psychiatry* **2023**, *28*, 5251–5261. [CrossRef] [PubMed]
160. Marie-Claire, C.; Lejeune, F.X.; Mundwiller, E.; Ulveling, D.; Moszer, I.; Bellivier, F.; Etain, B. A DNA Methylation Signature Discriminates between Excellent and Non-Response to Lithium in Patients with Bipolar Disorder Type 1. *Sci. Rep.* **2020**, *10*, 12239. [CrossRef]
161. Marie-Claire, C.; Courtin, C.; Bellivier, F.; Scott, J.; Etain, B. Methylomic Biomarkers of Lithium Response in Bipolar Disorder: A Proof of Transferability Study. *Pharmaceuticals* **2022**, *15*, 133. [CrossRef] [PubMed]
162. Zafrilla-López, M.; Acosta-Díez, M.; Mitjans, M.; Giménez-Palomo, A.; Saiz, P.A.; Barrot-Feixat, C.; Jiménez, E.; Papiol, S.; Ruiz, V.; Gavín, P.; et al. Lithium Response in Bipolar Disorder: Epigenome-Wide DNA Methylation Signatures and Epigenetic Aging. *Eur. Neuropsychopharmacol.* **2024**, *85*, 23–31. [CrossRef]
163. Brown, K.M.; Tracy, D.K. Lithium: The Pharmacodynamic Actions of the Amazing Ion. *Ther. Adv. Psychopharmacol.* **2013**, *3*, 163–176. [CrossRef] [PubMed]
164. Malhi, G.S.; Outhred, T. Therapeutic Mechanisms of Lithium in Bipolar Disorder: Recent Advances and Current Understanding. *CNS Drugs* **2016**, *30*, 931–949. [CrossRef] [PubMed]
165. Kato, T. Current Understanding of Bipolar Disorder: Toward Integration of Biological Basis and Treatment Strategies. *Psychiatry Clin. Neurosci.* **2019**, *73*, 526–540. [CrossRef] [PubMed]
166. Guilliot, S.; Gauthier, S.; Touchon, J.; Soto, M.E. Lithium, a Treatment Option for Alzheimer's Disease? A Review of Existing Evidence and Discussion on Future Perspectives. *J. Alzheimers Dis.* **2023**, *96*, 473–482. [CrossRef]
167. Adli, M.; Hollinde, D.L.; Stamm, T.; Wiethoff, K.; Tsahuridu, M.; Kirchheiner, J.; Heinz, A.; Bauer, M. Response to Lithium Augmentation in Depression Is Associated with the Glycogen Synthase Kinase 3-Beta-50T/C Single Nucleotide Polymorphism. *Biol. Psychiatry* **2007**, *62*, 1295–1302. [CrossRef]
168. Karati, D.; Meur, S.; Roy, S.; Mukherjee, S.; Debnath, B.; Jha, S.K.; Sarkar, B.K.; Naskar, S.; Ghosh, P. Glycogen Synthase Kinase 3 (GSK3) Inhibition: A Potential Therapeutic Strategy for Alzheimer's Disease. *Naunyn-Schmiedeberg Arch. Pharmacol.* **2024**, *2024*, 1–24. [CrossRef] [PubMed]
169. Singulani, M.P.; De Paula, V.J.R.; Forlenza, O.V. Mitochondrial Dysfunction in Alzheimer's Disease: Therapeutic Implications of Lithium. *Neurosci. Lett.* **2021**, *760*, 136078. [CrossRef]

pharmaceuticals

MDPI

Review

What Patients with Bipolar Disorder Need to Know about Lithium

Robert M. Post [1,*] and Janusz K. Rybakowski [2]

1 Bipolar Collaborative Network, Chevy Chase, MD 20815, USA
2 Department of Adult Psychiatry, Poznan University of Medical Sciences, 61-701 Poznań, Poland; janusz.rybakowski@gmail.com
* Correspondence: robert.postmd@gmail.com

Abstract: Lithium is the superior first-line treatment for bipolar disorder (BD). Yet the percentage of patients receiving lithium is abysmally low, especially in the US. Since psychiatrists have failed to place lithium in its appropriate role, we make the case that patients with BD themselves need to be better educated about the unique characteristics and pre-eminence of the drug so that it can be used more often and appropriately. Lithium has a highly unfavorable popular reputation among would-be patients and many psychiatrists. Thus, a direct appeal to patients with BD appears appropriate to try to remediate this situation. The unique assets of lithium are underappreciated or not well known. Conversely, the side effects profile of lithium are overestimated. Here, we make the case that lithium's image needs to be revised not only with better and more accurate information but also with a wholesale renaming and rebranding of the drug. We will not only outline the unique qualities and new information about the side effects of the drug but attempt to change some of the terminology conventionally used to refer to lithium so that its use may be appropriately applied earlier and at an increased frequency for patients with BD.

Keywords: lithium; bipolar disorder; therapeutics

1. Introduction

Lithium is the acknowledged superior first-line treatment for bipolar disorder [1]. Yet the percentage of patients receiving lithium is abysmally low, especially in the United States. Baldessarini et al. [2] found that the first class of drugs prescribed for bipolar patients was antidepressants in 50% and lithium in only 8%. Rhee et al. [3] found that in the last decade and a half, the use of antipsychotics increased from 12.4% to 51.4%, while the use of lithium decreased from 30.4% to 17.6%. Besides the US, the decline of lithium use in recent years was observed in Europe but not in Asia [4]. Elsewhere, we have argued that lithium should be reconsidered not only as a first-line drug, but also as a potential disease-modifying drug (DMD) [5], and as such, it should be used earlier and more often to reduce disability and dysfunction.

Lithium is unparalleled in its ability to prevent episodes of mania and depression [6,7]. It also has the best data for preventing suicides. It is most effective when initiated early in the course of bipolar disorder, and when used in this fashion has a major ameliorative effect on the series of adversities that are directly linked to the experience of increasing numbers of episodes [8]. These include an increased frequency and more rapid rate of recurrences; greater severity of recurrences; and episodes requiring fewer precipitating stressors. These sensitization and kindling-like processes emphasize the importance of early and sustained use of the best agents for preventing episodes. The underutilization of lithium is particularly problematic in the United States (US) where there is a higher incidence of childhood-onset bipolar disorder, and one-quarter of bipolar disorder in well-diagnosed adults begins before age 13 and two-thirds begins before age 19 [9–11]. These earlier onsets in the US are associated with greater illness severity as manifested by more childhood stressors; more anxiety disorders and substance abuse comorbidities; more episodes and rapid cycling; and greater degrees of treatment resistance during prospective longitudinal follow-up than in the Netherlands and Germany, abbreviated as Europe [12]. The earlier age of onset appears related to an increase in the incidence of genetic vulnerability for bipolar illness and psychiatric illness in general in the parents and grandparents of the patients from the US compared to Europe. In addition, there is also an increase in the incidence of childhood stressors (verbal, physical, and sexual abuse) in the US compared to Europe. Both of these genetic and environmental vulnerabilities are associated with an earlier age of onset of bipolar disorder and a more adverse course of illness in the US. It is noteworthy that the occurrence of verbal abuse alone (without the often accompanying physical or sexual abuse) is sufficient to be associated with the earlier age of onset and a more severe course of illness. As such, family-focused therapy (FFT) or another related family-based psychotherapy is a crucial component to the treatment of the precursors and earliest manifestations of the illness. This kind of therapy in addition to medications is associated with less depression and better outcomes, especially in families with highly expressed emotion [13,14].

Lithium is both less frequently used and started later in the course of illness in the US than in Europe. This could be a contributing factor in addition to the genetic and environmental vulnerabilities to the more adverse course of illness in the US. In addition to the antimanic effects of lithium demonstrated in controlled clinical trials, long-term follow-up data of Geller et al. [15] and Hafemen et al. [16] suggest that the use of lithium in youth with bipolar disorder is associated with a better outcome than with other agents and there is less depression and suicidal ideation when lithium is used compared with anticonvulsants and atypical antipsychotics. Thus, greater use of lithium in both childhood- and adult-onset bipolar disorder may help change the trajectory of illness into a more benign and less disabling form.

2. The Unique Assets of Lithium

Lithium not only prevents episodes but also prevents all of the indices of illness progression or what has been referred to as sensitization-like effects or the tendency of the illness to worsen as a function of the number of prior recurrences [8,17]. In addition to greater numbers and severity of episodes, more episodes are associated with greater cognitive impairment and treatment resistance. These sensitization effects are mediated by epigenetic effects based on chemical elements added to DNA and histones and changes in microRNA that alter how readily or not genes are expressed. Many of the epigenetic marks on sperm and ova survive after fertilization and alter vulnerability to illness in the next generation. Thus, in addition to the greater genetic vulnerability in the US compared to Europe, there is additional epigenetic vulnerability from the increases in parental stressors, episodes, and bouts of substance abuse that are also more prevalent in the US [12,17]. In addition, there are three times as much assortative mating (where two parents with mood disorders marry each other) in the US than in Europe, thus leaving more children in the US with a bilineal genetic as well as epigenetic vulnerability for psychiatric illness.

Therefore, given the high risk for childhood-onset bipolar illness in the US (7–9) and its relationship to a more adverse course of illness, extra measures should be taken to attempt to head off the multiple and progressive abnormalities associated with an increasing number of episodes (6, 13). As such, lithium should be given much higher consideration for early use in heading off episodes and illness progression [9–11,15,16]. In the case of medical illnesses such as multiple sclerosis (MS) [18] and rheumatoid arthritis (RA) [19], when disease-modifying drugs (DMDs) are used early in the course of these illnesses, they prevent the associated anatomical progression and functional deterioration accompanying these illnesses. Patients with these illnesses understand that delaying appropriate treatment with DMD drugs places them on a trajectory of irredeemable anatomical abnormalities and functional deficits. Given the many attributes of lithium that would appear to merit its designation as a disease-modifying drug (DMD) [5], earlier use may also help head off the neurobiological, anatomical, and functional deterioration that typically accompanies bipolar illness.

Parents with bipolar illness need to know that their children are at high risk for bipolar disorder and its related comorbidities such as ADHD, anxiety disorders, oppositional defiant disorder, and substance abuse [12,20]. Yet childhood-onset illness in the US is poorly recognized and treated. In addition to the diagnosis of early onset bipolar disorder itself, the delay to first treatment is longer in the US than in Europe and this delay is an independent predictor of more depression and a poor outcome in adulthood [21]. As such, parents with bipolar illness can assist clinicians and physicians in the identification and progression of impairing symptoms in their children.

An Institutional Review Board-approved system is available for parents to each week rate the severity of anxiety, depression, ADHD, oppositional behavior, and mania in their child so that the ratings can be printed out and shown to physicians in order to more systematically view illness course and response to treatment [22]. This system can be accessed at www.bipolarnews.org, accessed on 8 August 2024 (click on Child Network). These weekly parental ratings can readily display the range of inadequately treated symptoms in a child and help in the development and assessment of more effective treatment regimens.

Lack of and delayed use of lithium should be viewed from much the same perspective as that in MS or RA. Many of the deficits associated with multiple recurrences of episodes are not readily reversible or reversible at all. This is the case when one considers the many other fundamental assets of lithium in addressing the mechanism underlying the pathophysiology of the illness and illness progression. These deserve to be briefly enumerated here.

Lithium increases gray and white matter volume in the brain, and it is unique in increasing the volume of the hippocampus. It does this by increasing several neuroprotective factors BDNF and Bcl-2 and inhibiting cell death factors such as BAX and P-53 [23]. Thus, describing lithium in popular literature as toxic is a complete misnomer. It possesses the opposite characteristics of increasing cell survival. Lithium increases neurogenesis; it reduces all-cause mortality and extends longevity [24]. These are not trivial outcomes, as patients with bipolar disorder normally lose more than a decade of life expectancy, primarily based on increases in cardiovascular events, such as strokes and heart attacks [25]. Lithium also has the strongest data for preventing suicides which occur at a higher rate in bipolar disorder than in any other psychiatric illness [26].

Lithium increases the length of the end strands of DNA called telomeres, which shorten with depression, stress, and aging and are associated with increases in physical and psychiatric illnesses and comorbidities. Lithium activates the enzyme telomerase which places more DNA on the ends of telomeres [27] and does so in proportion to the duration of time a patient is treated with lithium [28].

Lithium also helps resolve some of the fundamental abnormalities associated with bipolar disorder, such as altered circadian rhythms. In addition, it has immunomodulatory effects and even anti-inflammatory and anti-viral effects [7].

While lithium has these unique properties, it is a too common misconception that if treatment with lithium leads to long-term remission, it is possible to slowly taper the drug and discontinue it. The data indicate that stopping lithium in the face of a good long-term response does not lessen the risk of a high incidence of relapse, about 50% within 5 months off the drug [29]. In a very small percentage of patients, there is an additional risk of not responding as well as they did before (or even at all) when treatment with lithium is re-instituted [30]. The first author knows about a dozen patients who have experienced this grievous outcome, and they all wanted others to know that stopping a treatment that has been working well in the long term, is not without considerable risk. This includes a severe relapse, suicide, or inducing an illness that has progressed to the next stage of lithium non-responsiveness and even refractoriness to treatment with multiple other agents as well.

3. Lithium's Adverse Effects Have Been Generally Over-Emphasized

One of the most widely touted side effects of lithium has been its causing kidney damage that can lead to the necessity of dialysis. The International Group for The Study of Lithium Treated Patients (IGSLI) evaluated lithium's impact on kidneys in 312 patients treated with lithium for 8–48 years (mean 18 years). It was concluded that long-term lithium treatment was associated with a gradual decline of renal functioning by about 30% more than that due to aging alone. The GFR declined by 0.7%/year of age and 0.9%/year of treatment, both by 19% more among women than men [31]. However, a very large nationwide study in Denmark has revealed that patients treated with lithium are no more likely to develop end-stage renal failure than those treated with anticonvulsants such as valproate or lamotrigine [32]. Moreover, recent changes in the way lithium is administered further decrease the likelihood of renal damage. This includes giving lithium in once-nightly doses, attempting to keep levels in the lower end (0.6–0.8 meq/L) of the therapeutic range (0.6 to 1.1 meq/L), and avoidance of episodes of lithium toxicity.

A more common side effect of lithium on the kidneys is lithium's ability to block the effects of the antidiuretic hormone vasopressin resulting in increases in urination called diabetes insipidus (DI). If urine amounts become problematic, DI can usually be well managed with the addition of the diuretic amiloride which markedly cuts the urine volume and frequency [33].

One often sees the potentially alarming statement that lithium is toxic to the thyroid. This disparaging and misleading characterization distorts the explanation that lithium can interfere with the production of thyroid hormones T3 and T4 and induce hypothyroidism in some 15 to 20% of patients. However, this deficit in hormone production can readily be alleviated with the replacement of thyroid hormone T4 [33].

Another common side effect of lithium is tremors, but in most cases, these can be avoided by lowering the dose of lithium. If this is not effective, the administration of propranolol at doses of 20–80 mg/day is a potentially effective strategy [33]. In cases where the patient is highly sensitive to lithium tremors, and an adequate dose cannot be achieved, lithium can be supplemented with the dihydropyridine calcium channel blocker nimodipine which has lithium-like effects without its tremor or other side effects [34,35]. Nimodipine specifically blocks the calcium influx gene CACNA1C which has been repeatedly linked to an increase in vulnerability to bipolar disorder and depression.

Weight gain is often erroneously touted as an invariable and problematic side effect of lithium. However, recent large long-term follow-up studies suggest that on average this is not the case [36].

Another widely held belief is that lithium impairs cognition or creativity. On the contrary, there is considerable evidence that over one year compared to placebo lithium in low doses prevents cognitive decline in those with mild cognitive impairment [37] and bipolar patients who use lithium have a lower incidence of dementia in old age than nonusers [38].

There are a variety of other less common potential side effects of lithium, but most can be dealt with by appropriate adjuncts or alternatives.

4. Questions That Patients with Bipolar Disorder Can Ask Their Physicians (Especially If They Are Not Already Being Treated with Lithium)

(1) Why am I not on the best medicine for treating and preventing bipolar illness progression?
(2) I have had one manic episode; is it not time to use lithium to head off further episodes and their long-term adverse consequences, including more rapid, severe, and untriggered relapses; cognitive dysfunction; disability; and ultimately treatment nonresponsiveness?
(3) I am on these other medications but still having episodes. Is it not time to add in lithium as it enhances the effectiveness of most other agents?
(4) If you as a physician or nurse practitioner are not well versed in the management of patients on lithium, can you refer me for a second opinion to an expert who is that I can see periodically?

(5) Bipolar illness is largely a progressive illness, is it not time that I used the best preventative medicine for it?

5. Some Counterintuitive Statements about Lithium

(1) Lithium is not a toxic drug in its usual doses. It is neuroprotective and reparative [7,39]. It increases the amount of gray matter and white matter in the brain, increases the volume of the hippocampus, and increases longevity.
(2) Since lithium likely qualifies as a disease-modifying drug (DMD), its use should be mandated earlier and more frequently in preventing the progression of bipolar illness [40].
(3) Bipolar illness is one of the psychiatric illnesses with the highest suicide rate, but lithium has the best data for preventing suicide [27]. There are also 10 studies indicating that micro-doses of lithium in drinking water are associated with lower rates of suicide in the general population [41].
(4) Lithium is an essential vitamin/mineral. Without its presence, normal growth and development do not occur. Li+ in lithium carbonate is an ion that has many similarities to the commonly used salt Na+ in sodium chloride, readily crosses cell membranes, and enters most cells in the brain and body. There are a myriad of potential mechanisms of action of lithium, but the ion has the exceptional properties of preventing or reversing most of the severe consequences of bipolar disorder. Calling Li+ the awesome ion or the incredible ion would not appear inappropriate.
(5) Contrary to the even lower utilization of lithium in bipolar disorder in children compared to adults, lithium is highly effective in childhood mania [42], and in long-term follow-up studies, children on lithium have better outcomes, including more periods of remission and less depression and suicide [15,16].
(6) Lithium in low doses prevents deterioration of mild cognitive impairment [37], and in therapeutic doses reduces the incidence of dementia in old age [38].
(7) There is almost no reason not to consider the use of lithium in patients with bipolar disorder. It has an extraordinary range of assets beyond its anti-manic effects (see Table 1 below).
(8) Most investigators now agree that prophylactic treatment of bipolar disorder should begin after a first mania, as two studies have shown that the recurrence of new episodes in the first year is associated with a lack of normalization of the deficits in cognition that occur compared to those with no further episodes.

The multiple assets of lithium beyond its anti-manic effect which were demonstrated in recent years are depicted in Table 1.

Table 1. Multiple assets of lithium beyond its anti-manic effects that patients should consider.

Variable	Data
Diagnosis	Lithium prevents unipolar and bipolar depression.
Suicide	Lithium prevents suicides in patients and in the general population.
	Lithium prevents suicide in the general population at minuscule doses in the water supply.
Augmentation	Lithium enhances the effects of other mood stabilizers and atypical antipsychotics.
Hippocampus	Lithium increases the volume of the hippocampus.
Gray/white matter	Lithium prevents or reverses cortical gray matter and white matter tract deficits.
Circadian rhythm	Lithium normalizes circadian rhythm abnormalities and has immune-modulating effects.
Immune system	Lithium has immune-modulating effects.
Telomeres	Lithium increases the length of telomeres shortened by episodes, stress, and aging.
Memory	Lithium prevents or slows memory deterioration in mild cognitive impairment over one year.

Table 1. *Cont.*

Variable	Data
Dementia	Lithium reduces the incidence of a diagnosis of dementia in old age.
Life expectancy	Lithium prolongs life expectancy and reduces the incidence of all-cause mortality.
All-cause mortality	Lithium reduces the incidence of all-cause mortality

6. Avenues for Future Research

A key area for future research would be the conduct of randomized controlled clinical trials of lithium in those at high risk for bipolar disorder by a parent with a diagnosis of bipolar disorder and in those with prodromal symptoms meeting criteria for bipolar not otherwise specified and related early manifestations of illness. Randomized clinical trials of lithium compared to other agents in early-onset bipolar disorder would also be of importance. For example, Berk et al. [43] reported that after a first hospitalization for mania, randomization to one year of treatment with lithium was superior to one year of quetiapine on all measures of mood, functioning, and abnormalities on brain imaging.

Clinical predictors of a good response to lithium include those with a BP I diagnosis with distinct episodes of euphoric mania, a lack of anxiety and substance abuse comorbidity, fewer episodes and less rapid cycling, and a positive family history of bipolar disorder and a good response to lithium. Valid neurobiological predictors of a good response to lithium remain to be ascertained and brought into clinical practice. When lithium is insufficient in producing a remission, further research is needed to more definitively assess what are the next best steps. This will be of considerable importance as lithium is able to enhance the efficacy of most other agents, including the mood-stabilizing anticonvulsants (lamotrigine, carbamazepine, and valproate); the dihydropyridine calcium channel blocker nimodipine; and virtually all of the atypical antipsychotics [44,45]. Only a select few AAs have definitive antidepressant effects in patients with bipolar depression. These include quetiapine in adults, lurasidone in children aged 10–17 and in adults, cariprazine, and lumateperone, but not aripiprazole. Lurasidone in both children and adults treated for bipolar depression is of particular interest because of the drug's good tolerability and weight neutrality and the unusual findings that the drug has the largest effect size in patients who have evidence of inflammation (high level of CRP) at baseline.

Another area for future study would be how to best address the multiple comorbidities that frequently accompany bipolar disorder but are less responsive to lithium including a wide range of anxiety and substance abuse comorbidities, eating disorders, and compulsions such as trichotillomania and nonsuicidal self-injury. These would be important areas for further study, especially since some commonly used supplements and other agents appear to show promising results. One of these N-acetylcysteine (NAC) has shown positive results in stabilizing mood and anxiety, in multiple types of addictions, obsessive compulsive disorder, trichotillomania, and other pathological habits [44,45]. Another is acetyl-L-carnitine which is low in the blood of patients with refractory depression (especially those with early onsets and a history of childhood adversity). It has good effects in depressed patients and in the defeat-stress animal model of depression where it works more rapidly than conventional antidepressants and appears to do so by an epigenetic mechanism of inducing an inhibitory glutamate receptor which decreases glutamate release [44,45]. The role of folate and L-methylfolate also deserves further exploration as enhancers of lithium and other antidepressant effects.

The promising area of further research on lithium is also that of pharmacogenetics, as indicated by the recent paper of Adiukwu et al. [46]. The gene variations that influence response to lithium were studied by the "candidate gene" method and more recently by the genome-wide association studies (GWASs). In 2021, two excellent reviews of the pharmacogenetics of lithium response were published [47,48].

7. Conclusions

Patients with multiple sclerosis (MS) and rheumatoid arthritis (RA) who do not use disease-modifying drugs (DMD) early in the course of their illness know they risk incurring irreversible anatomical and functional deficits. In parallel, patients with bipolar illness who do not use lithium which might also be called a DMD, risk incurring irreversible anatomical and functional deficits, including cognitive dysfunction, disability, dementia, and premature loss of life expectancy. However, some caution must be exercised since functional and morphological measurements in MS and RA, even if not perfect, are easier to estimate than behavioral, emotional, or neuromorphological changes in bipolar illness.

Bipolar disorder as it is conventionally treated in the United States is associated with a poor prognosis. Reversing the unwarranted neglect of lithium therapy in a very large segment of the patient population has the chance of making bipolar disorder a much less pernicious illness. The 75th anniversary of the introduction of lithium should be an occasion for the reassessment of the appropriate role of this drug in the treatment of the many adversities of bipolar disorder and reversing the many decades of its underutilization [49].

Funding: This research received no external funding.

Conflicts of Interest: The authors declare no conflicts of interest.

References

1. Malhi, G.S.; Bauer, M. Lithium first: Not merely first line. *Bipolar Disord.* **2023**, *25*, 7–8. [CrossRef] [PubMed]
2. Baldessarini, R.J.; Leahy, L.; Arcona, S.; Gause, D.; Zhang, W.; Hennen, J. Patterns of psychotropic drug prescription for U.S. patients with diagnoses of bipolar disorders. *Psychiatry Serv.* **2007**, *58*, 85–91. [CrossRef] [PubMed]
3. Rhee, T.G.; Olfson, M.; Nierenberg, A.A.; Wilkinson, S.T. 20-year trends in the pharmacologic treatment of bipolar disorder by psychiatrists in outpatient care settings. *Am. J. Psychiatry* **2020**, *177*, 706–715. [CrossRef] [PubMed]
4. Shuy, Y.K.; Santharan, S.; Chew, Q.H.; Sim, K. International trends in lithium use for pharmacotherapy and clinical correlates in bipolar disorder: A scoping review. *Brain Sci.* **2024**, *14*, 102. [CrossRef]
5. Post, R.M.; Li, V.; Yatham, L.N.; Berk, M. Is lithium a Disease Modifying Drug (DMD) for bipolar disorder? *Lancet Psychiatry* **2024**, *for submission*.
6. Kessing, L.V. Lithium as the drug of choice for maintenance treatment in bipolar disorder. *Acta Psychiatry Scand.* **2019**, *140*, 91–93. [CrossRef]
7. Rybakowski, J. Lithium treatment-the state of the art for 2020. *Psychiatry Pol.* **2020**, *54*, 1047–1066. [CrossRef]
8. Kessing, L.V.; Andersen, P.K. Evidence for clinical progression of unipolar and bipolar disorders. *Acta Psychiatry Scand.* **2017**, *135*, 51–64. [CrossRef]
9. Bellivier, F.; Etain, B.; Malafosse, A.; Henry, C.; Kahn, J.P.; Elgrabli-Wajsbrot, O.; Jamain, S.; Azorin, J.M.; Frank, E.; Scott, J.; et al. Age at onset in bipolar I affective disorder in the USA and Europe. *World J. Biol. Psychiatry* **2011**, *15*, 369–376. [CrossRef]
10. Etain, B.; Lajnef, M.; Bellivier, F.; Mathieu, F.; Raust, A.; Cochet, B.; Gard, S.; M'Bailara, K.; Kahn, J.P.; Elgrabli, O.; et al. Clinical expression of bipolar disorder type I as a function of age and polarity at onset: Convergent findings in samples from France and the United States. *J. Clin. Psychiatry* **2012**, *73*, e561–e566. [CrossRef] [PubMed]
11. Post, R.M. Parents with mood disorders should know the risks for their children. *Prev. Med. Commun. Health* **2017**, *1*, 1–4. [CrossRef]
12. Post, R.M.; Altshuler, L.L.; Kupka, R.; McElroy, S.L.; Frye, M.A.; Rowe, M.; Grunze, H.; Suppes, T.; Keck, P.E., Jr.; Leverich, G.S.; et al. More childhood onset bipolar disorder in the United States than Canada or Europe: Implications for treatment and prevention. *Neurosci. Biobehav. Rev.* **2017**, *74*, 204–213. [CrossRef] [PubMed]
13. Miklowitz, D.J.; Schneck, C.D.; Singh, M.K.; Taylor, D.O.; George, E.L.; Cosgrove, V.E.; Howe, M.E.; Dickinson, L.M.; Garber, J.; Chang, K.D. Early intervention for symptomatic youth at risk for bipolar disorder: A randomized trial of family-focused therapy. *J. Am. Acad. Child. Adolesc. Psychiatry* **2013**, *52*, 121–131. [CrossRef] [PubMed]
14. Miklowitz, D.J.; Merranko, J.A.; Weintraub, M.J.; Walshaw, P.D.; Singh, M.K.; Chang, K.D.; Schneck, C.D. Effects of family-focused therapy on suicidal ideation and behavior in youth at high risk for bipolar disorder. *J. Affect. Disord.* **2020**, *275*, 14–22. [CrossRef]
15. Geller, B.; Tillman, R.; Bolhofner, K.; Zimerman, B. Pharmacological and non-drug treatment of child bipolar I disorder during prospective eight-year follow-up. *Bipolar Disord.* **2010**, *12*, 164–171. [CrossRef]
16. Hafeman, D.L.; Rooks, B.; Merranko, J.; Liao, F.; Gill, M.K.; Goldstein, T.R.; Diler, R.; Ryan, N.; Goldstein, B.I.; Axelson, D.A.; et al. Lithium versus other mood-stabilizing medications in a longitudinal study of youth diagnosed with bipolar disorder. *J. Am. Acad. Child. Adolesc. Psychiatry* **2020**, *59*, 1146–1155. [CrossRef]
17. Post, R.M. Epigenetic basis of sensitization to stress, affective episodes, and stimulants: Implications for illness progression and prevention. *Bipolar Disord.* **2016**, *18*, 315–324. [CrossRef] [PubMed]

18. Iaffaldano, P.; Portaccio, E.; Lucisano, G.; Simone, M.; Manni, A.; Guerra, T.; Paolicelli, D.; Betti, M.; De Meo, E.; Pastò, L.; et al. Italian Multiple Sclerosis Register. Multiple sclerosis progression and relapse activity in children. *JAMA Neurol.* **2024**, *81*, 50–58. [CrossRef]
19. Benjamin, O.; Goyal, A.; Lappin, S.L. Disease-Modifying Antirheumatic Drugs (DMARD. In *StatPearls [Internet]. Treasure Island (FL)*; StatPearls Publishing: St. Petersburg, FL, USA, 2024. [PubMed]
20. Post, R.M.; Altshuler, L.L.; Kupka, R.; McElroy, S.L.; Frye, M.A.; Rowe, M.; Grunze, H.; Suppes, T.; Keck, P.E.; Leverich, G.S.; et al. More illness in offspring of bipolar patients from the U.S. compared to Europe. *J. Affect. Disord.* **2016**, *191*, 180–1866. [CrossRef] [PubMed]
21. Post, R.M.; Leverich, G.S.; Kupka, R.W.; Keck, P.E., Jr.; McElroy, S.L.; Altshuler, L.L.; Frye, M.A.; Luckenbaugh, D.A.; Rowe, M.; Grunze, H.; et al. Early-onset bipolar disorder and treatment delay are risk factors for poor outcome in adulthood. *J. Clin. Psychiatry* **2010**, *71*, 864–872. [CrossRef]
22. Post, R.M.; Rowe, M.; Kaplan, D.; Findling, R.L. A multisymptomatic child with bipolar disorder: How to track and sequence treatment. *Prim. Care Companion CNS Disord.* **2017**, *19*, 17r02122. [CrossRef]
23. Quiroz, J.A.; Machado-Vieira, R.; Zarate, C.A., Jr.; Manji, H.K. Novel insights into lithium's mechanism of action: Neurotrophic and neuroprotective effects. *Neuropsychobiology* **2010**, *62*, 50–60. [CrossRef] [PubMed]
24. Carvalho, A.F.; Hsu, C.W.; Vieta, E.; Solmi, M.; Marx, W.; Berk, M.; Liang, C.S.; Tseng, P.T.; Wang, L.J. Mortality and lithium-protective effects after first-episode mania diagnosis in bipolar disorder: A nationwide retrospective cohort study in Taiwan. *Psychother Psychosom.* **2024**, *93*, 36–45. [CrossRef] [PubMed]
25. Chan, J.K.N.; Tong, C.H.Y.; Wong, C.S.M.; Chen, E.Y.H.; Chang, W.C. Life expectancy and years of potential life lost in bipolar disorder: Systematic review and meta-analysis. *Br. J. Psychiatry* **2022**, *221*, 567–576. [CrossRef] [PubMed]
26. Tondo, L.; Vázquez, G.H.; Baldessarini, R.J. Prevention of suicidal behavior in bipolar disorder. *Focus* **2023**, *21*, 402–411. [CrossRef]
27. Martinsson, L.; Wei, Y.; Xu, D.; Melas, P.A.; Mathe, A.A.; Schalling, M.; Lavebratt, C.; Backlund, L. Long-term lithium treatment in bipolar disorder is associated with longer leukocyte telomeres. *Transl. Psychiatry* **2013**, *3*, e261. [CrossRef]
28. Squassina, A.; Meloni, A.; Congiu, D.; Bosganas, P.; Patrinos, G.P.; Lin, R.; Turecki, G.; Severino, G.; Ardau, R.; Chillotti, C.; et al. Analysis on in vitro effect of lithium on telomere length in lymphoblastoid cell lines from bipolar disorder patients with different clinical response to long-term lithium treatment. *Hum. Genom.* **2022**, *16*, 45. [CrossRef]
29. Suppes, T.; Baldessarini, R.J.; Faedda, G.L.; Tohen, M. Risk of recurrence following discontinuation of lithium treatment in bipolar disorder. *Arch. Gen. Psychiatry* **1991**, *48*, 1082–1088. [CrossRef]
30. Post, R.M. Acquired lithium resistance revisited: Discontinuation-induced refractoriness versus tolerance. *J. Affect. Disord.* **2012**, *140*, 6–13. [CrossRef]
31. Tondo, L.; Abramowicz, M.; Alda, M.; Bauer, M.; Bocchetta, A.; Bolzani, L.; Calkin, C.V.; Chillotti, C.; Hidalgo-Mazzei, D.; Manchia, M.; et al. Long-term lithium treatment in bipolar disorder: Effects on glomerular filtration rate and other metabolic parameters. *Int. J. Bipolar Disord.* **2017**, *5*, 27. [CrossRef]
32. Kessing, L.V.; Gerds, T.A.; Feldt-Rasmussen, B.; Andersen, P.K.; Licht, R.W. Use of lithium and anticonvulsants and the rate of chronic kidney disease: A nationwide population-based study. *JAMA Psychiatry* **2015**, *72*, 1182–1191. [CrossRef]
33. Ferensztajn-Rochowiak, E.; Rybakowski, J.K. Long-term lithium therapy: Side effects and interactions. *Pharmaceuticals* **2023**, *16*, 74. [CrossRef] [PubMed]
34. Pazzaglia, P.J.; Post, R.M.; Ketter, T.A.; George, M.S.; Marangell, L.B. Preliminary controlled trial of nimodipine in ultra-rapid cycling affective dysregulation. *Psychiatry Res.* 1993; 49, 257–272.
35. Pazzaglia, P.J.; Post, R.M.; Ketter, T.A.; Callahan, A.M.; Marangell, L.B.; Frye, M.A.; George, M.S.; Kimbrell, T.A.; Leverich, G.S.; Cora-Locatelli, G.; et al. Nimodipine monotherapy and carbamazepine augmentation in patients with refractory recurrent affective illness. *J. Clin. Psychopharmacol.* **1998**, *18*, 404–413. [CrossRef] [PubMed]
36. Gomes-da-Costa, S.; Marx, W.; Corponi, F.; Anmella, G.; Murru, A.; Pons-Cabrera, M.T.; Giménez-Palomo, A.; Gutiérrez-Arango, F.; Llach, C.D.; Fico, G.; et al. Lithium therapy and weight change in people with bipolar disorder: A systematic review and meta-analysis. *Neurosci. Biobehav. Rev.* **2022**, *134*, 104266. [CrossRef] [PubMed]
37. Forlenza, O.V.; Diniz, B.S.; Radanovic, M.; Santos, F.S.; Talib, L.L.; Gattaz, W.F. Disease-modifying properties of long-term lithium treatment for amnestic mild cognitive impairment: Randomised controlled trial. *Br. J. Psychiatry* **2011**, *198*, 351–356. [CrossRef]
38. Kessing, L.V.; Forman, J.L.; Andersen, P.K. Does lithium protect against dementia? *Bipolar Disord.* **2010**, *12*, 87–94. [CrossRef]
39. Rowe, M.K.; Chuang, D.M. Lithium neuroprotection: Molecular mechanisms and clinical implications. *Expert Rev. Mol. Med.* **2004**, *6*, 1–18. [CrossRef]
40. Kessing, L.V.; Vradi, E.; Andersen, P.K. Starting lithium prophylaxis early v. late in bipolar disorder. *Br. J. Psychiatry* **2014**, *205*, 214–220. [CrossRef]
41. Strawbridge, R.; Young, A.H. Lithium: How low can you go? *Int. J. Bipolar Disord.* **2024**, *12*, 4. [CrossRef]
42. Findling, R.L.; Robb, A.; McNamara, N.K.; Pavuluri, M.N.; Kafantaris, V.; Scheffer, R.; Frazier, J.A.; Rynn, M.; DelBello, M.; Kowatch, R.A.; et al. Lithium in the acute treatment of bipolar I disorder: A double-blind, placebo-controlled study. *Pediatrics* **2015**, *136*, 885–894. [CrossRef]
43. Berk, M.; Daglas, R.; Dandash, O.; Yucel, M.; Henry, L.; Hallam, K.; Macneil, C.; Hasty, M.; Pantelis, C.; Murphy, B.P.; et al. Quetiapine v. lithium in the maintenance phase following a first episode of mania: Randomized controlled trial. *Br. J. Psychiaty* **2017**, *210*, 413–442. [CrossRef]

44. Post, R.M.; Yatham, L.N.; Vieta, E.; Berk, M.; Nierenberg, A. Beyond evidence-based treatment of bipolar disorder: Rationa pragmatic approaches to management. *Bipolar Disord.* **2019**, *21*, 650–659. [CrossRef] [PubMed]
45. Post, R.M.; Goldstein, B.; Birmaher, B.; Findling, R.L.; Frey, B.; DelBello, M.; Miklowitz, D. Toward prevention of bipolar disorder in at-risk children: Potential strategies ahead of the data. *J. Affect. Disord.* **2020**, *272*, 508–520. [CrossRef] [PubMed]
46. Adiukwu, F.; Adesokun, O.; Essien, E.; Yalcin, N.; Ransing, R.; Nagendrappa, S.; Jatchavala, C.; Olakunke, A.B.; Nawaz, F.A.; Khan, N.A. Pharmacogenetic testing in psychiatry: Perspective on clinical utility. *Asian J. Psychiatry* **2023**, *86*, 103674.
47. Senner, F.; Kohshour, M.O.; Abdalla, S.; Papiol, S.; Schulze, T.G. The genetics of response to and side effects of lithium treatment in bipolar disorder: Future research perspectives. *Front. Pharmacol.* **2021**, *12*, 638882. [CrossRef] [PubMed]
48. Vecera, C.M.; Fries, G.R.; Shahani, L.R.; Soares, J.C.; Machado-Vieira, R. Pharmacogenomics of lithium response in bipolar disorder. *Pharmaceuticals* **2021**, *14*, 287. [CrossRef]
49. Post, R.M. The new news about lithium: An underutilized treatment in the United States. *Neuropsychopharmacology* **2018**, *43*, 1174–1179. [CrossRef]

pharmaceuticals

MDPI

Protocol

Lithium Point-of-Care Testing to Improve Adherence to Monitoring Guidelines and Quality of Maintenance Therapy: Protocol for a Randomised Feasibility Trial

Jess Kerr-Gaffney [1,*], Priyanka Prakash [1], Victoria C. Wing [2,3,4], Allan H. Young [1,5], Oisín N. Kavanagh [6], John Hodsoll [7], Sarah Markham [7], David A. Cousins [2,3,4], Elliot Hampsey [1], Sameer Jauhar [1,5], David Taylor [8], Anthony J. Cleare [1] and Rebecca Strawbridge [1]

1 Department of Psychological Medicine, Institute of Psychiatry, Psychology & Neuroscience, King's College London, London SE5 8AB, UK; becci.strawbridge@kcl.ac.uk
2 Newcastle Magnetic Resonance Centre, Newcastle University, Health Innovation Neighbourhood, Newcastle upon Tyne NE4 5PL, UK
3 Cumbria, Northumberland, Tyne and Wear NHS Foundation Trust, Newcastle upon Tyne NE3 3XT, UK
4 Mental Health and Neurodegeneration and Dementia Theme, Translational and Clinical Research Institute, Newcastle University, Newcastle upon Tyne NE1 7RU, UK
5 Division of Psychiatry, Imperial College London, London SW7 2AZ, UK
6 School of Pharmacy, Newcastle University, Newcastle upon Tyne NE1 7RU, UK; oisin.kavanagh@newcastle.ac.uk
7 Department of Biostatistics & Health Informatics, Institute of Psychiatry, Psychology & Neuroscience, King's College London, London WC2R 2LS, UK
8 Institute of Pharmaceutical Science, King's College London, Franklin Wilkins Building, London SE1 9NH, UK
* Correspondence: jess.kerr-gaffney@kcl.ac.uk

Abstract: Lithium is the first-line treatment for bipolar disorders and a first-line augmentation option for treatment-resistant unipolar depression. Due to its narrow therapeutic window and risk of toxicity, people taking lithium require regular blood testing to monitor lithium levels in the body. However, studies have reported that only half of lithium-treated patients receive adequate lithium monitoring. This protocol describes a trial that will test the feasibility and acceptability of a point-of-care (POC) lithium blood testing programme in patients with unipolar or bipolar affective disorders taking lithium as a maintenance treatment. The primary objectives are to establish whether testing the effectiveness of POC testing is feasible, by assessing recruitment, attrition, and adherence to monitoring guidelines compared to participants randomised to testing as usual; to test whether the programme is acceptable to patients; and to measure potential contamination bias. The secondary objectives are to examine changes in health-related quality of life, the use of healthcare services, and depressive and manic symptoms to inform the design of a larger multi-site randomised controlled trial (RCT). This feasibility RCT will recruit 80 participants with affective disorders who are taking lithium. Participants will be 1:1 randomised to either POC monitoring or monitoring as usual where they will be followed up at three research visits over 30 weeks. The proportion of patients meeting guidelines for lithium monitoring will be examined, alongside measures of acceptability, wellbeing, and health economic data. POC testing has the potential to significantly improve patient safety and satisfaction with lithium treatment.

Keywords: lithium; protocol; randomised controlled trial; bipolar; depression; point of care

1. Introduction

Lithium, a mood stabiliser, is the recommended first-line long-term pharmacological treatment for bipolar disorder (BD) [1] and is used as an augmentation treatment for those with unipolar depression who have not responded to an initial course of antidepressants [2]. Due to its narrow therapeutic window and risk of toxicity, people taking lithium require regular blood testing to monitor lithium levels throughout the course of treatment. The National Institute for Health and Care Excellence (NICE) recommends that lithium concentrations should range between 0.6 and 0.8 mmol/L for those who are being prescribed lithium for bipolar disorder for the first time or up to 1.0 mmol/L for those who have previously relapsed or continue to experience symptoms whilst receiving lithium [1]. Those with lithium levels below the therapeutic window may be at increased risk of relapse, whilst those with levels above the therapeutic window are at increased risk of side effects and lithium toxicity [3]. This is problematic as lithium levels are highly variable and are influenced by the timing of the lithium dose and factors such as food and water intake, exercise, and concomitant medications [4,5].

NICE guidelines recommend that lithium levels be monitored every 3 months in the first year of treatment and every 6 months thereafter, although high-risk groups (e.g., older adults, those taking drugs which interact with lithium, or people at risk of impaired renal or thyroid function) should continue to be monitored every 3 months [1]. Levels should also be measured 1 week after any dose changes. However, audit data from NHS services within the UK reported that only 30% of patients met monitoring guidelines for the frequency of lithium testing in 2009 [6]. Following a national quality improvement programme [7], this rose to 48% [8]; however, half of patients continued to receive inadequate lithium monitoring. A later study in one NHS trust reported that only 50.7% of patients were within the NICE-recommended therapeutic lithium range, meaning that many patients may not be receiving effective symptom relief or are at risk of more severe side effects [9]. Studies outside of the UK have reported similar findings [10,11]. Various procedural and individual-level factors may contribute to poor monitoring rates, including a lack of communication between services and differing practitioner roles.

Currently, lithium testing is undertaken via laboratory testing, which typically involves patients attending a service specifically to have venous blood drawn by a nurse or phlebotomist, and the test results are then added to a patient's health record within a few days. However, a number of inefficiencies have been identified in this system, including the need to book a separate blood test, with the appointment often being scheduled weeks after being ordered by the doctor; patients having to call practices on busy phone lines for their test results; non-clinical staff communicating test results; and a lack of formal protocols for communicating results [12,13]. The perceived burden of blood testing has been reported to contribute to the declining rates of lithium prescribing worldwide [14,15].

This study aims to test the feasibility of a novel, point-of-care (POC) lithium blood testing programme to improve monitoring in people with unipolar and bipolar affective disorder. POC testing allows for lithium serum levels to be measured by members of the patient's treatment team during an appointment using finger prick testing, giving results in minutes. POC testing has revolutionised other areas of healthcare, for example, glucose monitoring in diabetes [16]. Recently, POC testing was introduced to one NHS trust for clozapine monitoring in schizophrenia. The new system was found to be highly acceptable to patients and staff: 90% of patients felt that the system improved clinical care, and staff felt that the system allowed for faster changes to prescription, the identification of non-compliance, and ease of administration [17]. The initiative received two (of six) findings of outstanding practice from the Care Quality Commission Inspection [18] and is now used routinely within several NHS trusts. Introducing POC testing into psychiatry may help bring mental health monitoring in line with physical health practices, such as those used in diabetes management.

The primary objective of this study is to test the feasibility and acceptability of POC lithium blood testing compared to usual monitoring for patients with affective disorders (bipolar disorder and unipolar depression) receiving lithium maintenance treatment, in a randomised controlled trial (RCT) design. Feasibility will be assessed via participant recruitment and retention rates, lithium discontinuation rates, and adherence to monitoring guidelines. Contamination bias (e.g., crossover effects) will also be assessed. Acceptability will be assessed via clinician and patient questionnaires. The secondary objectives are to measure health-related outcomes, service utilisation, mood, adverse events, and side effects.

2. Methods

2.1. Design

The Lithium Point-of-Care testing (LiPOC) trial is a multi-site, two-arm, rater-blinded RCT investigating POC lithium testing versus testing as usual (TAU) over 30 weeks. A total of 80 participants will be recruited across four sites in London and one site in the North of England: South London and Maudsley NHS Foundation Trust (SLaM), West London NHS Trust (WLNHS), South West London & St George's NHS Trust (SWLSTG), Central and North West London NHS Trust (CNWL), and Cumbria, Northumberland, Tyne and Wear NHS Foundation Trust (CNTW). Participants will be randomised 1:1 to POC or TAU. Participants will remain under the care of their treating clinicians, meaning that their lithium testing will take place within their usual services, regardless of allocation.

2.2. Participants

Participants will be recruited from any service within the five participating NHS trusts. They may be identified via (a) direct research team communications with clinicians; (b) adverts within trust premises; (c) searches of the clinical records of patients who have consented to be contacted for research, such as SLaM's Consent for Contact and linked Clinical Records Interactive Search; and (d) existing databases of patients who, through other routes (such as previous study participation), have consented to be contacted by researchers within the Centre for Affective Disorders, King's College London, or CNTW.

Inclusion criteria are as follows: (a) aged 18 and above, (b) a diagnosis of a unipolar or bipolar affective disorder, (c) due to have monitoring for lithium treatment at the beginning of this study and three monthly for the upcoming 6-month period, and (d) be receiving care at one of the participating trusts. Exclusion criteria are as follows: (a) a planned discontinuation of lithium treatment during the 30-week study period; (b) an affective disorder diagnosis for which lithium is not licenced (i.e., schizoaffective disorder); (c) clinically significant manic symptoms, defined as a Young Mania Rating Scale (YMRS) [19] score ≥ 20; and (d) inability to provide informed consent.

We will also invite clinicians to participate in an optional add-on study at the end of their patient's participation in the trial to assess clinician experiences with the intervention. The only inclusion criterion is that the clinician must have been involved in the POC intervention of their patient who was enrolled in our trial.

2.3. Procedure

A study flowchart is shown in Figure 1. Participants will attend three visits: at week 0 (baseline), week 15, and week 30. All visits can take place remotely (i.e., via video call) or in person. Potential participants will first attend a screening visit to assess their eligibility and obtain verbal consent. Eligible participants will then be invited to attend a baseline visit, where written informed consent will be obtained, and demographic and baseline assessments will be administered (see Table 1). Following this, participants will be randomised, and the intervention will begin.

Follow-up visits will involve clinical and acceptability questionnaires, and any medication changes will be documented. In addition, researchers will obtain the levels of lithium, thyroid function test results, and serum levels of sodium, calcium, and creatinine via medical records. A timeline summarising patient visits during this study is shown in Figure 2.

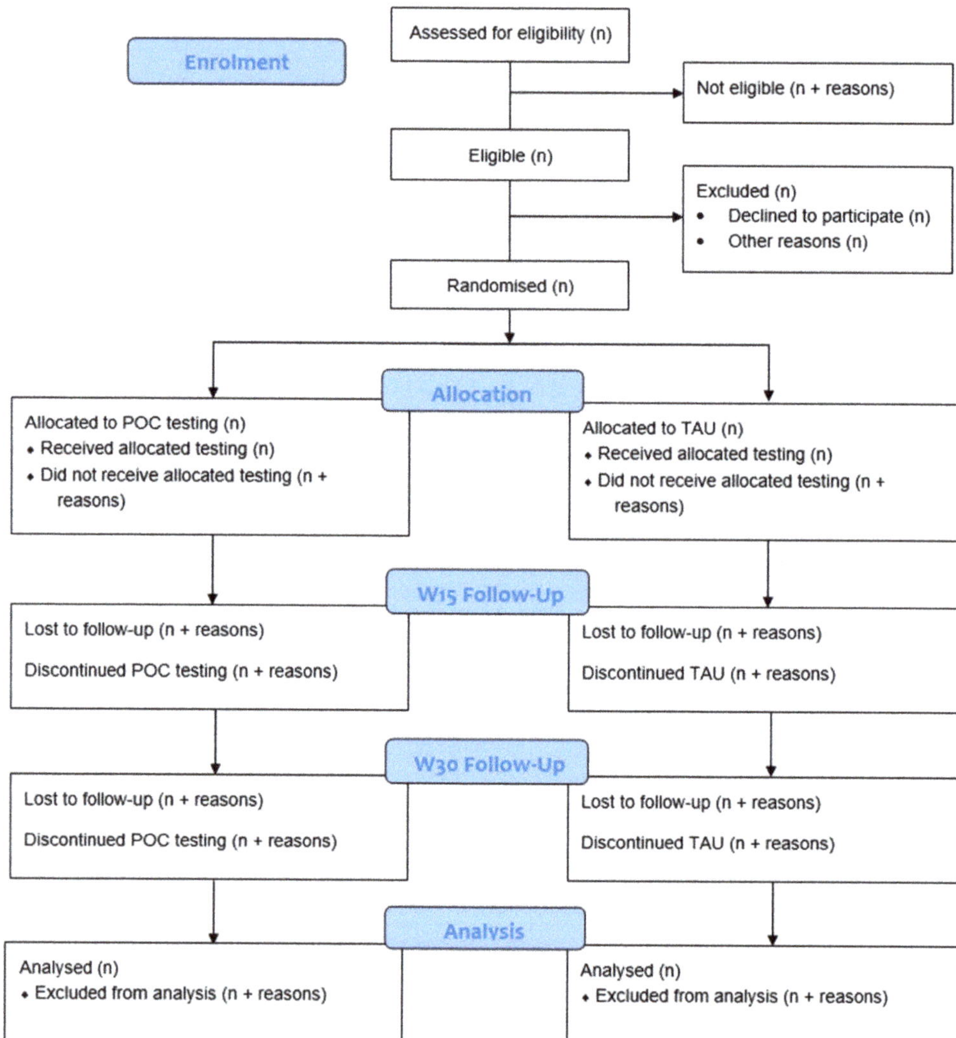

Figure 1. Study flowchart.

Table 1. Study procedures by visit.

Procedures/Measures	Screening (Pre-Study)	Baseline (W0)	Follow-Up 1 (W15)	Follow-Up 2 (W30)
Informed consent process	X	X		
Medical and psychiatric information	X	X		
Sociodemographic information	X	X		
Medication and lithium review	X	X	X	X
Randomisation		X		
EuroQol-5D (EQ-5D)		X	X	X
Client Service Receipt Inventory (CSRI)		X	X	X
Young Mania Rating Scale (YMRS)	X	X	X	X
Montgomery–Åsberg Depression Rating Scale (MADRS)		X	X	X
Maudsley visual analogue scales (M3VAS)		X	X	X
Lithium side-effects rating scale (LiSERS)		X	X	X
Patient monitoring acceptability questionnaire		X	X	X
Participant end of study feedback questions				X

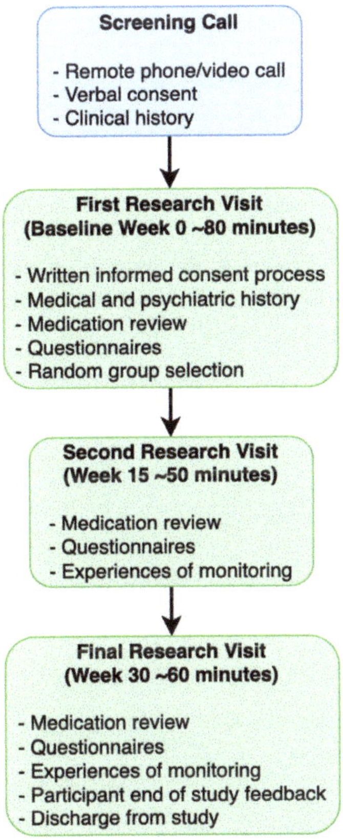

Figure 2. Summary of patient visits and assessments. Blue indicates screening (pre-study), green indicates baseline and follow-up period.

2.4. Intervention

2.4.1. POC Testing

Participants randomised to POC testing will have their lithium and creatinine serum levels tested with the Medimate device, a handheld tool which provides serum lithium and creatinine concentration from a finger prick blood sample within a few minutes. The device has been tested in line with international protocols, i.e., Clinical Laboratory Standards Institute Evaluation Protocols, which facilitate validation for manufacturers and laboratories in the US and EU. The device demonstrates an Allowable Total Error (ATE) rate for lithium levels below 10% or 0.1 mmol/L, exceeding both US CLIA regulations (maximum 20% or 0.3 mmol/L) and European consensus standards, and has a testing capacity of 0.2 mmol/L to 12 mmol/L [20]. It is also validated for measuring creatinine. Participants' healthcare professionals will be asked to notify the study team when and where they are arranging the participants' lithium test so that a study researcher can bring the testing device to administer the test to those in the POC arm. In this group, the Medimate device will be used for all lithium monitoring from randomisation up until week 30 of this study (provided agreement from clinicians and that they do not discontinue prior to this).

2.4.2. TAU

Participants randomised to TAU will not receive any change to testing and will continue via laboratory testing as usual.

2.5. Randomisation and Blinding

Participants will be randomised 1:1 to POC or TAU within 24 h of completing the baseline assessment using King's College Clinical Trials Unit independent web-based system. Randomisation will be stratified by site. The participant and healthcare professional will be notified of their allocation as soon as possible.

Participants and their clinicians will be aware of their treatment allocation. Researchers conducting the YMRS and Montgomery–Åsberg Depression Rating Scale (MADRS) outcome assessments will be blinded and will not have any other roles in the trial to maintain blinding. The trial statistician will also be blinded. The database will only refer to groups 01 and 02 without definition to maintain the blinding of the statistician.

2.6. Outcomes

Primary outcomes are as follows:

1. Feasibility
    a. Participant recruitment rates over the trial period.
    b. Participant retention and completion rates over 30 weeks.
    c. Rates of lithium discontinuation over 30 weeks.
    d. Adherence to lithium monitoring guidelines (proportion of participants meeting NICE guidelines for frequency testing and lithium levels within therapeutic range) over 30 weeks.
2. Patient acceptability
    a. Patient monitoring acceptability questionnaires at week 15 and week 30.
    b. Patient preferences for POC testing versus laboratory testing.
    c. Perceived usefulness of POC monitoring.
3. Contamination bias
    a. Monitoring rates between treatment-exposed versus non-exposed services.
    b. Changes in monitoring rates from first to last patients recruited.

c. Participants' monitoring rates from before trial (from electronic health records) versus during the trial.
d. Clinician survey about service changes during the trial.

Secondary outcomes are as follows:

1. Costs and economic outcomes
    a. Health related-quality of life via the EuroQol-5D (EQ-5D) [21] at weeks 15 and 30.
    b. Service utilisation via the Client Service Receipt Inventory (CSRI) [22] at weeks 15 and 30.
2. Mood and affective symptoms
    a. Manic symptoms via the YMRS at weeks 15 and 30 [19].
    b. Depression symptoms via the MADRS [23] at weeks 15 and 30.
    c. Depression, mania, and anxiety symptoms via the Maudsley visual analogue scales (M3VAS) [24] at weeks 15 and 30.
3. Side effects and adverse events
    a. Lithium side-effects rating scale (LiSERS) [25] at weeks 15 and 30.
    b. Adverse events over 30 weeks.

2.7. Statistics

The sample size of 80 will enable the estimation of a 50% dropout rate with a confidence interval of ±10% [26].

Variables will be summarised using descriptive statistics (i.e., mean and standard deviation [SD] or median and interquartile range [IQR] or frequencies and proportions, as appropriate). Recruitment, retention, discontinuation, and measure completion rates will be described. The number and proportion of participants meeting NICE guidelines [1] for test frequency and therapeutic range compliance (usually 0.6 mmol/L–0.8 mmol/L but in certain specified circumstances either 0.8 mmol/L–1.0 mmol/L or 0.4–0.6 mmol/L) will be reported between study arms. Because our aim is to take three measurements per participant within the 30-week trial period, we are recruiting patients who require 3-monthly monitoring (with measurements at baseline/week 0, week 15, and week 30). However, due to the practical timing of clinic visits, some participants may only complete two measurements within this period; therefore two or three measurements will be considered adherent to monitoring guidelines, as long as they align with the participant's scheduled monitoring timeframes. Acceptability questionnaires will be described quantitatively within subjects and between groups over time, as well as a qualitative analysis of narrative response questions.

The following will be analysed descriptively to investigate potential contamination bias (i.e., mean and SD or median and IQR or frequencies and proportions, as appropriate): (a) monitoring rates between treatment-exposed vs. non-exposed services, (b) changes in monitoring rates from the first to last patients recruited, (c) participants' monitoring history from electronic health records, and (d) ratings from surveying clinicians about practice changes during the trial.

The unit costs of the intervention and use of other services will be estimated with the CSRI and compared between groups to identify the main cost impacts that should be measured in a future trial, in combination with health-related quality of life data from the EQ-5D to estimate quality-adjusted life years (QALYs). Generalised linear mixed models will estimate the between-arm differences at 15 and 30 weeks for the M3VAS, YMRS, MADRS, and LiSERS. These analyses will be used to estimate effect sizes for a future definitive RCT, as this feasibility RCT is not powered to detect treatment differences. The generalised linear mixed models will use all available data under a maximum likelihood framework. Missing follow-up data will not be imputed. Adverse events will be summarised descriptively between arms.

Analyses will be intention to treat (ITT). Sensitivity analyses may be carried out, excluding participants without clear information on their lithium treatment or monitoring.

Table 2 details key feasibility criteria to be met to proceed with a larger definitive RCT.

Table 2. Go–no-go criteria to be met to proceed with an efficacy trial.

	Go—Proceed with RCT	Amend—Proceed with Changes	Stop—Do Not Proceed Unless Changes Are Possible
Participant recruitment	80 in 12 months	80 reached but in >12 months	53 not reached by 12 months
POC monitoring acceptability	≥80% rating POC as "acceptable"	30–79% rating POC as "acceptable"	<30% rating POC as "acceptable"
Participant attrition rate	≤20% attrition	21–50% attrition	>50% attrition
Adherence to lithium monitoring guidelines	>60% adherence	30–60% adherence	<30% adherence

2.8. Trial Oversight

The trial received a favourable opinion from the London—South East Research Ethics Committee (25/LO/0381). This study will be conducted in compliance with the principles of the Declaration of Helsinki (1996) and Good Clinical Practice (GCP). King's College London and South London and Maudsley NHS Foundation Trust are the sponsors of this trial. The trial was registered with the ISRCTN registry (ISRCTN17989376, on 27 May 2025). Participant recruitment commenced in September 2025 and will continue until September 2026.

A Trial Steering Committee (TSC) will provide oversight throughout the duration of the trial. The TSC aims to ensure the trial is being conducted to GCP and other regulatory standards, that participant welfare is being considered in the design and running of the trial, and that sufficient progress is being made with regard to key milestones (e.g., participant recruitment).

2.9. Data Security and Management

The chief investigator will act as custodian for the study data. All study data will be stored in line with the Data Protection Act and archived in line with sponsor requirements. The source data for the study will be collected on paper case report forms (CRFs). Data will then be input into the trial database, which will be hosted on a secure server hosted within King's College London. Access to the trial database will be restricted through user-specific passwords to authorised research team members. The database will only store pseudonymised data, not including any direct patient identifiers (e.g., name, date of birth, or contact information). Unique participant identification numbers (PINs) will be used within the database and paper CRFs. Paper forms will be stored securely within locked cabinets at each site.

The study team will undertake reviews of the entered data where appropriate for the purpose of data cleaning and will request amendments as required. Following checks of data correctness and completeness, all data will be formally locked for analysis.

2.10. Patient and Public Involvement (PPI)

A lived experience advisory panel (LEAP) comprising individuals with experience of unipolar or bipolar depression and lithium treatment will meet throughout the trial, to provide insight into the real-world impact of the trial activities, processes, and findings.

The trial team also includes a PPI lead, who will be part of the TMG and lead the LEAP and co-authored this protocol. Patient-facing documents were reviewed by the Feasibility and Acceptability Support Team for Researchers (FAST-R), a team of people with lived experience of mental health problems and their carers who advise on research proposals and documentation.

2.11. Dissemination

The results of the trial will be reported in peer-reviewed scientific journals and presented at conferences. A primary publication will include all primary and secondary outcomes listed in the protocol. Further publications may report exploratory analyses. The results will also be disseminated to the public via plain English summaries, with input from the LEAP.

3. Discussion

This protocol describes a feasibility RCT which aims to test the utility of a novel POC testing method for lithium monitoring in the UK. Data gathered on feasibility, acceptability, and adherence to monitoring guidelines will be used to inform a larger confirmatory trial, which would establish whether POC testing is clinically effective and cost-effective. If POC testing is found to be effective in improving lithium monitoring in those with unipolar and bipolar affective disorders, this could lead to several benefits for this population. If there is an improvement in the proportion of time that participants' lithium levels are kept within the therapeutic range, this could improve affective symptoms and reduce the side effects of lithium, enhancing quality of life. Further, POC testing may lead to direct and indirect cost benefits, for example, by eliminating the need for a separate appointment for a blood test and therefore reducing service use requirements and patient time off work. Further, prior research using POC testing for those taking antipsychotics has indicated that patients felt more involved in their own care and that receiving the test results almost immediately helped them understand the purpose of testing [17]. This could lead to improved engagement, trust, and communication between patients and their healthcare team.

This trial has a number of strengths and limitations. The trial will use a novel drug level monitoring method that has been shown to be acceptable and preferred over laboratory testing in a similar patient group [17]. Further, although this is a feasibility trial, the inclusion of five NHS sites across two regions of England will help improve the generalisability of the results, ensuring that the results are not an artefact of processes at one NHS trust. Although double-blinding is not possible in this trial (patients and at least some of the patient's healthcare team would need to know their allocation), blinded-outcome raters will complete assessments for key clinical outcomes with patients to reduce bias. Related to this, it is possible that clinicians who are notified that their patient is taking part in a lithium monitoring trial will take more care to follow clinical guidelines for lithium monitoring. Measures to reduce contamination bias will be taken; for example, when patients' clinicians are notified of their involvement in the trial, minimal detail on the intervention will be given to clinicians whose patients are in the TAU arm to reduce the chance that they will change their monitoring practices. They are told that the patient is taking part in a trial relating to lithium treatment but not that the trial relates to lithium monitoring. Further, a multi-method assessment of contamination bias is a primary objective of this trial, incorporating analyses of monitoring rates from patients before they entered the trial, comparing monitoring rates from exposed versus non-exposed services, changes in monitoring rates across the recruitment period, and feedback from clinicians.

Because this is a feasibility trial, inferences about efficacy cannot be made from the results. A further limitation of this trial is that we are only including patients who require 3-monthly monitoring, therefore excluding a large proportion of patients on lithium maintenance treatment who require monitoring every 6 months. This decision was made in order to have an adequate number of scheduled monitoring appointments per participant within the time constraints of a feasibility trial. If this trial is successful, we would hope to also include those on 6-month monitoring in the larger confirmatory trial that would follow, which may include a longer follow-up period. A 12-month follow-up period would allow for the inclusion of patients who require 3-monthly or 6-monthly monitoring, with 4 or 2 expected monitoring appointments within the trial follow-up period, respectively. This would also allow longer-term adherence and efficacy to be assessed. However, a 12-month follow-up period presents challenges with regard to research investment. Finally, although the POC testing method provides lithium and creatinine serum levels, it does not provide monitoring of other parameters that the clinician may require or are recommended while taking lithium, such as urea and electrolytes, calcium, and thyroid function. NICE guidelines recommend that these parameters be monitored every 6 months or more frequently in certain circumstances [1].

The quality of lithium monitoring in the UK and elsewhere is currently poor, potentially contributing to poor health outcomes in those with unipolar and bipolar affective disorders. Known issues with current blood monitoring practices may lead to patients being outside the therapeutic range for lithium, putting them at risk of relapse or sub-optimal symptom management if levels are too low or compromising physical health if levels are too high. The LiPOC trial aims to build on previous work introducing POC testing to psychiatry, with the hope of improving care and quality of life in this population. Further, despite lithium being the most efficacious drug for preventing mood episodes in bipolar disorder, the rates of prescribing are declining [27]. If POC testing is eventually adopted in healthcare services and is perceived by clinicians and patients as less burdensome and more efficient than laboratory testing, clinicians may be more likely to prescribe lithium to their patients, providing more patients with the chance to trial an effective treatment for their condition.

Author Contributions: Conceptualisation: A.H.Y., R.S., J.K.-G. and D.T. Methodology: R.S., J.K.-G., A.H.Y. and J.H. Investigation: J.K.-G., P.P., V.C.W., O.N.K., S.M., D.A.C., E.H., S.J., A.J.C. and R.S. Writing—Original Draft Preparation: J.K.-G. Writing—Review and Editing: all authors. Funding Acquisition: R.S., J.K.-G., A.H.Y., J.H. and S.M. All authors have read and agreed to the published version of the manuscript.

Funding: The trial described in this protocol paper is funded by a National Institute for Health and Care Research (NIHR) Research for Patient Benefit grant (NIHR207148).

Institutional Review Board Statement: The trial received a favourable opinion from the London–South East Research Ethics Committee (25/LO/0381; approval date: 4 June 2025).

Informed Consent Statement: Informed consent is obtained from all participants involved in the study.

Data Availability Statement: No new data were created or analyzed in this study. Data sharing is not applicable to this article.

Acknowledgments: We would like to thank the NIHR Maudsley BRC FAST-R service for their input into the patient-facing documents for this trial. We would also like to thank the NIHR for funding and peer review support; Eric Lynch, Mitul Mehta, Robyn Wilcha, Ryan Williams, the NIHR Research Delivery Network, BipolarUK, our Lived Experience Advisory Panel, supporting clinicians, services, and researchers within our participating services; and our study participants.

Conflicts of Interest: J.K.-G. owns shares in AstraZeneca and GSK. R.S. has received honoraria for educational sessions from Janssen. A.J.C. has received grant funding from the MRC, ADM Protexin Ltd., NIHR, European Union Horizon Europe/Innovate UK, Beckley Psytech Ltd., and Wellcome Trust; has received honoraria for presentations and/or consulting from Janssen, Otsuka, COMPASS Pathways Plc, Viatris and Medscape; and is President of the International Society for Affective Disorders. A.H.Y. reports the following: employed by Imperial College London; Honorary Consultant Psychiatrist (NHS UK); Editor of Journal of Psychopharmacology and Deputy Editor, BJPsych Open; Paid lectures and advisory boards for the following companies with drugs used in affective and related disorders: Flow Neuroscience, Novartis, Roche, Janssen, Takeda, Noema pharma, Compass, Astrazenaca, Boehringer Ingelheim, Eli Lilly, LivaNova, Lundbeck, Sunovion, Servier, Livanova, Janssen, Allegan, Bionomics, Sumitomo Dainippon Pharma, Sage, Neurocentrx; and grant funding (past and present): NIMH (USA); CIHR (Canada); NARSAD (USA); Stanley Medical Research Institute (USA); MRC (UK); Wellcome Trust (UK); Royal College of Physicians (Edin); BMA (UK); UBC-VGH Foundation (Canada); WEDC (Canada); CCS Depression Research Fund (Canada); MSFHR (Canada); NIHR (UK); Janssen (UK) EU Horizon 2020. S.J. has received honoraria for educational talks from Janssen, Lundbeck, Sunovian, and Boehringer Ingelheim and honoraria as an adviser for LB pharmaceuticals Inc. He is a Council Member for the British Association for Psychopharmacology and sat as a panel member for the Wellcome Trust. D.T. has received investigator-initiated research grants from Janssen; received speaker fees from Lundbeck, Otsuka, Viatris, and Recordati; and has shares in Myogenes, Saladax, and 428-Pharma. J.H., O.N.K., S.M., D.A.C., P.P., E.H., and V.W. declare no competing interests.

Abbreviations

The following abbreviations are used in this manuscript:

ATE	Allowable Total Error
BD	Bipolar Disorder
CNTW	Cumbria, Northumberland, Tyne and Wear NHS Foundation Trust
CNWL	Central and North West London NHS Trust
CSRI	Client Service Receipt Inventory
EQ-5D	EuroQol-5D
FAST-R	Feasibility and Acceptability Support Team for Researchers
GCP	Good Clinical Practice
ITT	Intention to Treat
IQR	Interquartile Range
LEAP	Lived Experience Advisory Panel
LiSERS	Lithium Side-Effects Rating Scale
M3VAS	Maudsley Visual Analogue Scales
MADRS	Montgomery–Åsberg Depression Rating Scale
mmol/L	Millimoles Per Litre
NHS	National Health Service
NICE	National Institute for Health and Care Excellence
POC	Point of Care
RCT	Randomised Controlled Trial
SD	Standard Deviation
SLaM	South London and Maudsley NHS Foundation Trust
SWLSTG	South West London & St George's NHS Trust
TAU	Treatment As Usual
TSC	Trial Steering Committee
WLNHS	West London NHS Trust
YMRS	Young Mania Rating Scale

References

1. National Institute for Health and Care Excellence. Bipolar Disorder: Assessment and Management [CG185]. 2025. Available online: https://www.nice.org.uk/guidance/cg185 (accessed on 1 August 2025).
2. National Institute for Health and Care Excellence. Depression in Adults: Treatment and Management [NG222]. 2022. Available online: http://www.nice.org.uk/guidance/ng222 (accessed on 1 August 2025).
3. Severus, W.; Kleindienst, N.; Seemüller, F.; Frangou, S.; Möller, H.; Greil, W. What Is the Optimal Serum Lithium Level in the Long-term Treatment of Bipolar Disorder—A Review? *Bipolar Disord.* **2008**, *10*, 231–237. [CrossRef]
4. Gitlin, M. Lithium Side Effects and Toxicity: Prevalence and Management Strategies. *Int. J. Bipolar Disord.* **2016**, *4*, 27. [CrossRef]
5. Zorrilla, I.; Lopez-Zurbano, S.; Alberich, S.; Barbero, I.; Lopez-Pena, P.; García-Corres, E.; Chart Pascual, J.P.; Crespo, J.M.; De Dios, C.; Balanzá-Martínez, V.; et al. Lithium Levels and Lifestyle in Patients with Bipolar Disorder: A New Tool for Self-Management. *Int. J. Bipolar Disord.* **2023**, *11*, 11. [CrossRef]
6. Collins, N.; Barnes, T.R.; Shingleton-Smith, A.; Gerrett, D.; Paton, C. Standards of Lithium Monitoring in Mental Health Trusts in the UK. *BMC Psychiatry* **2010**, *10*, 80. [CrossRef] [PubMed]
7. National Patient Safety Agency. Safer Lithium Therapy [NPSA/2009/PSA005]. 2009. Available online: https://www.cas.mhra.gov.uk/ViewandAcknowledgment/ViewAlert.aspx?AlertID=101306 (accessed on 1 August 2025).
8. Paton, C.; Adroer, R.; Barnes, T.R. Monitoring Lithium Therapy: The Impact of a Quality Improvement Programme in the UK. *Bipolar Disord.* **2013**, *15*, 865–875. [CrossRef] [PubMed]
9. Nikolova, V.L.; Pattanaseri, K.; Hidalgo-Mazzei, D.; Taylor, D.; Young, A.H. Is Lithium Monitoring NICE? Lithium Monitoring in a UK Secondary Care Setting. *J. Psychopharmacol.* **2018**, *32*, 408–415. [CrossRef] [PubMed]
10. Uskur, T.; Güven, O.; Tat, M. Retrospective Analysis of Lithium Treatment: Examination of Blood Levels. *Front. Psychiatry* **2024**, *15*, 1414424. [CrossRef]
11. Nederlof, M.; Egberts, T.C.; Van Londen, L.; De Rotte, M.C.; Souverein, P.C.; Herings, R.M.; Heerdink, E.R. Compliance with the Guidelines for Laboratory Monitoring of Patients Treated with Lithium: A Retrospective Follow-up Study among Ambulatory Patients in the Netherlands. *Bipolar Disord.* **2019**, *21*, 419–427. [CrossRef]
12. Litchfield, I.; Bentham, L.; Hill, A.; McManus, R.J.; Lilford, R.; Greenfield, S. Routine failures in the process for blood testing and the communication of results to patients in primary care in the UK: A qualitative exploration of patient and provider perspectives. *BMJ Qual. Saf.* **2015**, *24*, 681–690. [CrossRef]
13. Litchfield, I.; Bentham, L.; Lilford, R.; McManus, R.J.; Hill, A.; Greenfield, S. Test result communication in primary care: A survey of current practice. *BMJ Qual. Saf.* **2015**, *24*, 691–699. [CrossRef]
14. Hidalgo-Mazzei, D.; Mantingh, T.; Pérez de Mendiola, X.; Samalin, L.; Undurraga, J.; Strejilevich, S.; Severus, E.; Bauer, M.; González-Pinto, A.; Nolen, W.A.; et al. Clinicians' preferences and attitudes towards the use of lithium in the maintenance treatment of bipolar disorders around the world: A survey from the ISBD Lithium task force. *Int. J. Bipolar Disord.* **2023**, *11*, 20. [CrossRef]
15. Malhi, G.S.; Bell, E.; Jadidi, M.; Gitlin, M.; Bauer, M. Countering the declining use of lithium therapy: A call to arms. *Int. J. Bipolar Disord.* **2023**, *11*, 30. [CrossRef]
16. Olansky, L.; Kennedy, L. Finger-Stick Glucose Monitoring. *Diabetes Care* **2010**, *33*, 948–949. [CrossRef]
17. Atkins, M.; Taylor, D.; Harland, R.; Brewer, A.; Williams, S.; Chesney, E.; McGuire, P. Acceptability of Point of Care Testing for Antipsychotic Medication Levels in schizophrenia. *Psychiatry Res. Commun.* **2022**, *2*, 100070. [CrossRef]
18. Care Quality Commission. South London and Maudsley NHS Foundation Trust: Inspection Report. 2021. Available online: https://api.cqc.org.uk/public/v1/reports/4a1743ed-a98a-4b9a-adfa-7bfa83639c77?20221129062700 (accessed on 1 August 2025).
19. Young, R.C.; Biggs, J.T.; Ziegler, V.E.; Meyer, D.A. A Rating Scale for Mania: Reliability, Validity and Sensitivity. *Br. J. Psychiatry* **1978**, *133*, 429–435. [CrossRef] [PubMed]
20. Staal, S.; Ungerer, M.; Floris, A.; Ten Brinke, H.; Helmhout, R.; Tellegen, M.; Janssen, K.; Karstens, E.; Van Arragon, C.; Lenk, S.; et al. A Versatile Electrophoresis-based Self-test Platform. *Electrophoresis* **2015**, *36*, 712–721. [CrossRef] [PubMed]
21. Herdman, M.; Gudex, C.; Lloyd, A.; Janssen, M.; Kind, P.; Parkin, D.; Bonsel, G.; Badia, X. Development and Preliminary Testing of the New Five-Level Version of EQ-5D (EQ-5D-5L). *Qual. Life Res.* **2011**, *20*, 1727–1736. [CrossRef]
22. Beecham, J.; Knapp, M. Costing Psychiatric Interventions. In *Measuring Mental Health Needs*; Gaskell: London, UK, 1999; pp. 200–224.
23. Montgomery, S.A.; Åsberg, M. A New Depression Scale Designed to Be Sensitive to Change. *Br. J. Psychiatry* **1979**, *134*, 382–389. [CrossRef] [PubMed]
24. Moulton, C.D.; Strawbridge, R.; Tsapekos, D.; Oprea, E.; Carter, B.; Hayes, C.; Cleare, A.J.; Marwood, L.; Mantingh, T.; Young, A.H. The Maudsley 3-Item Visual Analogue Scale (M3VAS): Validation of a Scale Measuring Core Symptoms of Depression. *J. Affect. Disord.* **2021**, *282*, 280–283. [CrossRef]
25. Haddad, P.; Wieck, A.; Yarrow, M.; Denham, P. The Lithium Side Effects Rating Scale (LISERS); Development of a Self-Rating Instrument. *Eur. Neuropsychopharmacol.* **1999**, *9*, 231–232. [CrossRef]

26. National Institute for Health and Care Research. Additional Guidance For Applicants Including a Clinical Trial, Pilot Study or Feasibility as Part of a Personal Award Application. 2019. Available online: https://www.nihr.ac.uk/additional-guidance-applicants-including-clinical-trial-pilot-study-or-feasibility-part-personal-award-application (accessed on 1 August 2025).
27. Kessing, L.V. Why is lithium [not] the drug of choice for bipolar disorder? A controversy between science and clinical practice. *Int. J. Bipolar Disord.* **2024**, *12*, 3. [CrossRef]

pharmaceuticals

MDPI

Review

Overlap and Divergence in Ketamine and Lithium Response in Bipolar Disorder: A Scoping Review

Jay Toulany [1,*], Jasmyn E. A. Cunningham [1] and Abraham Nunes [1,2]

[1] Department of Psychiatry, Faculty of Medicine, Dalhousie University, Halifax, NS B3H 2E2, Canada; nunes@dal.ca (A.N.)
[2] Faculty of Computer Science, Dalhousie University, Halifax, NS B3K 6R8, Canada
* Correspondence: jay.toulany@dal.ca

Abstract: Background/Objectives: Lithium remains the first choice for long-term prophylaxis of mood episodes in bipolar disorder (BD), but only 30% of patients will respond, and there is no reliable method by which to predict treatment response. Ketamine is a rapid antidepressant therapy which ostensibly yields greater results in patients with clinical phenotypes that are classically associated with lithium non-response. This inspired a scoping review to map the overlapping and divergent clinical and mechanistic evidence for acute ketamine response and long-term prophylactic lithium therapy in BD. **Methods:** We conducted a scoping review of clinical and preclinical studies that examine convergent and divergent predictors and mechanisms of acute response to ketamine and long-term response to lithium. **Results:** Data from 19 preclinical studies show mechanistic convergence of ketamine and lithium on the GSK-3β/mTOR pathways, and enhancement of synaptic plasticity. Furthermore, lithium appears to consistently limit ketamine-related oxidative stress and hyperlocomotion. However, data from the 23 clinical studies suggest divergence of predictors of response to ketamine and lithium in BD, with ketamine response associated with metabolic risk factors, anxiety/mixed features, and non-melancholic presentations, which are generally predictors of poorer prophylactic lithium response. No study directly tested ketamine response as a predictor of prophylactic lithium response. An important limitation is that clinical studies of ketamine are enriched for lithium-refractory populations and have often included mixed unipolar and bipolar cases. **Conclusions:** Overall, existing data support mechanistic overlap but clinical divergence between ketamine and lithium responders, though this is confounded by sampling bias. We must therefore undertake longitudinal studies of prophylactic lithium therapy among patients with BD who received ketamine for acute antidepressant treatment in order to investigate if responsiveness to ketamine predicts response to lithium, and establish control over BD earlier in the course of illness.

Keywords: bipolar disorder; depression; ketamine; lithium; treatment prediction; maintenance therapy

1. Introduction

Bipolar disorder (BD) is a chronic neuropsychiatric disorder for which effective long-term maintenance treatment is critical to preventing relapses and reducing morbidity [1]. After more than seven decades, lithium remains the gold-standard prophylactic mood stabilizer in BD [2]. However, only one-third of patients achieve an excellent response, which is characterized by full symptomatic remission and functional recovery [3,4]. Since an unsuccessful lithium trial can entail months of suboptimal treatment and side effects, psychiatrists and researchers must strive to find a priori or early markers of eventual response to lithium prophylaxis [5].

Decades of research suggest that patients with a more "classical" bipolar phenotype are the most likely to respond well to lithium maintenance [4,6,7]. Indeed, lithium responsiveness is associated with a completely episodic course (discrete episodes with full symptomatic remission and functional recovery in the inter-episode period), absence of rapid cycling, a strong family history of BD, and a later age of onset [4,5,8]. By contrast, lithium non-responsive BD patients typically exhibit a more atypical or "non-classical" clinical phenotype [6], including the presence of mixed features, rapid cycling, and significant comorbidities, such as anxiety disorders [4,6,9]. Other studies have noted that metabolic comorbidities, such as insulin resistance, diabetes, and higher body mass index (BMI), are more common in lithium non-responders [8,10]. Since clinicians currently cannot definitively determine lithium response before initiating therapy, many ultimately lithium non-responsive patients are subjected to lengthy lithium trials to determine effectiveness. These patients are at risk of experiencing many months of side-effects and illness recurrence if lithium eventually proves ineffective.

In recent years, ketamine (an NMDA receptor antagonist) has emerged as a novel augmentation option for acute treatment-resistant bipolar depression [11]. Sub-anesthetic doses of ketamine produce rapid antidepressant effects within hours to days, which compares favorably to the 4–8 weeks typically required for response to conventional antidepressants [12,13]. Intriguingly, ketamine appears to benefit some subgroups of patients that are traditionally difficult to treat, such as patients with non-melancholic or "anxious" depressive subtypes [14], or patients with significant comorbid mixed mood features [15,16]. Furthermore, a higher body mass index (BMI) has predicted positive response to ketamine [17]. Taken together, these findings suggest that acute ketamine responders may phenotypically resemble prophylactic lithium non-responders, implying that ketamine's mechanism of action may potentially target acute depressive pathology in patient populations whose clinical phenotypes diverge from those typically observed in responders to long-term lithium prophylaxis.

Therefore, we sought to understand whether non-response to an acute ketamine trial could serve as a positive predictor of long-term lithium efficacy, and whether a robust response to ketamine might predict a lower probability of lithium maintenance success through the evaluation of both clinical and preclinical evidence. The implications of such a relationship, if proven, would be highly significant. The rapid antidepressant effect of ketamine offers a unique opportunity to biologically "test" a patient's treatment responsiveness within days, whereas evaluating lithium's prophylactic effect requires months to years of observation. If a brief ketamine infusion series could reliably indicate whether a patient is more or less likely to respond to prophylactic lithium, the efficiency of long-term treatment decisions in BD could be greatly improved.

Up front, an important methodological consideration arises from the inclusion criteria of most clinical trials of ketamine in bipolar depression; specifically, that studying treatment-resistant samples will increase the probability that included patients had already failed to respond to multiple prior pharmacologic interventions, including mood stabilizers such as lithium. As a result, these study populations are likely enriched for individuals who are already known or presumed to be lithium non-responders. This could create a systematic sampling bias limiting the generalizability of findings and obscuring potential relationships between ketamine and lithium response. Thus, we considered it important to examine the overlapping and distinct mechanisms of action of ketamine and lithium in order to evaluate their common and differential effects in patients with BD. This paper therefore presents a scoping review of the clinical and preclinical evidence supporting or refuting this proposed link between ketamine response and lithium maintenance outcomes in BD.

2. Methods

This scoping review adheres to the Preferred Reporting Items for Systematic Review and Meta-Analysis for Scoping Reviews (PRISMA-ScR) guidelines (Figure 1) [18].

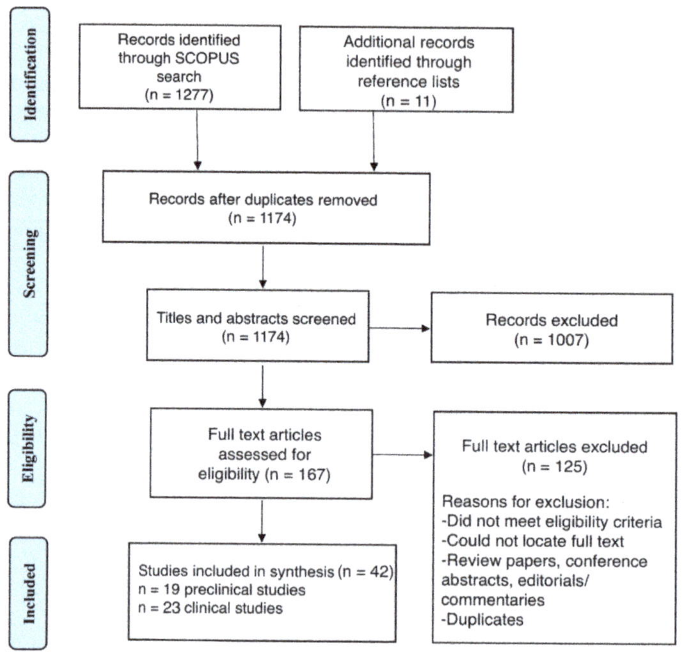

Figure 1. Preferred Reporting Items for Systematic Reviews and Meta-Analyses (PRISMA) flow diagram of study selection.

2.1. Data Collection

The Scopus database was searched from its inception to 26 April 2025, using the following search query:

TITLE-ABS-KEY (ketamine AND (lithium OR anticonvuls* OR antipsychot* OR mood-stabiliz* OR (mood PRE/0 stabiliz*)) AND (bipolar OR depress*)). Reviews, conference abstracts, commentaries, and editorials were excluded from the review but their reference lists were manually examined to identify additional potentially relevant publications for inclusion.

2.2. Screening Process

Two authors (J.T. and A.N.) independently assessed the titles and abstracts of the retrieved articles for potential inclusion. Disagreements between reviewers were resolved through follow-up discussion to reach a consensus. Inter-rater reliability was not formally assessed during screening. One author (J.T.) then assessed full-text papers published in English for final inclusion. Articles were included in the present review if they met the following criteria: (1) Clinical trials that investigated the use of ketamine and a mood stabilizer in the treatment of bipolar depression, and which reported depressive symptom severity ratings over time; (2) Studies that administered ketamine in combination with a mood stabilizer to animals; (3) Preclinical translational studies that explored the mechanistic relationship between ketamine and mood stabilizers, examining both overlapping and distinct mechanisms; (4) Studies that specifically evaluated clinical predictors of ketamine response. Unpublished data, reviews, commentaries, and editorial articles were not eligible for inclusion in the present review. We included studies where ketamine was administered with other prophylactic treatments, such as lamotrigine and clozapine, in order to further explore the mechanistic overlap and divergence between ketamine, lithium, and these other prophylactic agents.

For all studies, data extracted by J.T. and A.N. included study information (e.g., author, publication details), participant/animal population, study design, intervention details, and relevant depressive outcomes. Study data were then categorized, summarized, and qualitatively synthesized by the study team. Despite the limitations of preclinical studies in capturing the complexity of BD, relatively equal weighting was given to the data extracted from preclinical and clinical studies, as the mechanistic insights gained from preclinical evidence helped address our study aims.

3. Results

The initial search yielded a total of 1277 records, with 11 additional records identified from relevant reference lists (Figure 1). After removing 114 duplicates, 1174 records were screened for relevance based on titles and abstracts, with 1007 excluded at this stage. Following full-text review of 167 articles, 125 studies were excluded based on the eligibility criteria previously outlined. In total, 42 records that met the inclusion criteria were included in the qualitative synthesis, 19 of which were preclinical studies and 23 of which were clinical studies.

3.1. Preclinical Studies

3.1.1. Lithium Augmentation of the Rapid Antidepressant Effect of Ketamine

Chiu et al. [19] showed that a single administration of high-dose ketamine (50 mg/kg delivered intraperitoneally [i.p.]) reverses depressive-like behavior in chronically stressed mice (measured as a reduction in immobility in forced swim and tail suspension tests). However, this antidepressant effect is transient and accompanied by increased oxidative stress markers in the brain. Notably, low-dose lithium given before ketamine enabled a previously ineffective subthreshold ketamine dose (2.5 mg/kg) to produce antidepressant-like behavioral effects. Likewise, lithium delivered post-ketamine (at therapeutic levels) sustained ketamine's antidepressant action and synaptic enhancements for at least 2 weeks. Furthermore, Chiu et al. found that lithium pretreatment completely abolished ketamine-induced oxidative damage (lipid peroxidation, glutathione oxidation) in the stressed mouse brain [19]. Mechanistically, lithium-enhanced ketamine-mediated activation of the mTOR–BDNF pathway in the prefrontal cortex (PFC) resulted in increased dendritic spine density associated with antidepressant response. This suggests that combining ketamine with lithium can both potentiate the acute antidepressant effect and limit ketamine-related adverse oxidative stress.

Ketamine's antidepressant action may be related to GSK-3β inhibition and consequent increases in synaptic plasticity. Liu et al. [20] demonstrated that blocking GSK-3, either with lithium or a selective inhibitor, markedly potentiated ketamine's effects. In their study, a normally subthreshold ketamine dose produced robust antidepressant-like effects only when combined with a lithium/GSK-3 inhibitor, resulting in rapid mTORC1 activation, elevated BDNF–TrkB signaling, and increased density/strength of synaptic spines. The combination (low-dose ketamine + lithium) was as effective as a higher ketamine dose alone, highlighting GSK-3 as a mediator of ketamine's synaptogenic and behavioral effects. As a clinical corollary, this may also support a treatment strategy for patients who may not be able to tolerate normally therapeutic ketamine doses (i.e., potentiation with a low dose of lithium). In line with this, do Vale et al. [21] found that lithium synergized with ketamine to enhance antidepressant outcomes and anti-inflammatory signaling. Low-dose ketamine (2 mg/kg) had only modest effects on forced-swim immobility, but when paired with lithium it yielded a reduction in immobility time comparable to a 10 mg/kg ketamine dose. This combination also potentiated a ketamine-mediated reduction in pro-inflammatory markers (TNF-α, iNOS, COX-2) and increased inhibitory GSK-3β phosphorylation in peripheral tissue. These findings suggest that GSK-3 inhibition is a critical mechanism that boosts ketamine-driven antidepressant response and synaptic plasticity.

Recent data also implicate insulin–Akt (insulin-linked kinases) signaling in ketamine's sustained antidepressant effect. In a rodent model of treatment-resistant depression, Price et al. [22] showed that adding lithium to ketamine not only reduced immobility time and increased latency to immobility in the forced swim test, but also uniquely upregulated insulin pathway signaling. Rats receiving ketamine and lithium had significantly higher plasma insulin levels and mTOR activation, along with increased phosphorylation of Akt in the prefrontal cortex. Blood insulin levels correlated inversely with depression-like immobility in these animals receiving both ketamine and lithium, and increased insulin signaling in the infralimbic PFC also correlated with behavioral improvement. These results suggest lithium's augmentation may recruit metabolic signaling pathways to support antidepressant effects.

At the circuit level, ketamine's antidepressant action appears to be related to its acute psychotomimetic effects. For example, Maltbie et al. [23] used fMRI in awake nonhuman primates to map ketamine-induced brain activation. Ketamine infusion broadly increased BOLD activation, with the strongest activations in the cingulate gyrus, supplementary motor area (SMA), and thalamus, but this was reduced in both extent and magnitude by risperidone pretreatment. There was no deactivation of the subgenual cingulate observed—an area where overactivity has been implicated in treatment-resistant depression [24]. It is thus possible that disinhibition of PFC networks is a shared mechanism for the rapid antidepressant effect of ketamine and its transient psychotic symptoms. The lack of acute antipsychotic effects of lithium in humans may thus suggest that divergence in the effects of ketamine and lithium could relate to mechanisms related to psychotomimesis.

Further circuit-related findings suggest that ketamine and lithium may overlap in their effects on stress-related brain circuits. Specifically, in a study of rats by Stepan et al. [25] using hippocampal slice imaging, the authors found that chronic stress impairs propagation of neural activity through the hippocampus, whereas lithium, ketamine, and BDNF ameliorate this impairment. Such findings suggest that ketamine and lithium both reinforce synaptic connectivity in stress-sensitive circuits, further supporting convergent mechanistic targets for these agents.

3.1.2. Ketamine as a Preclinical Model of Mania/Psychosis

Repeated low-dose ketamine in rodents can produce hyperactivity and dysregulation thought to be analogous to manic episodes. This ketamine-induced "mania-like" behavior is typically characterized by increased locomotor activity, risk-taking, and dysregulated neurotransmitter signaling. A consistent finding across such studies is that ketamine triggers oxidative stress and inflammation in brain regions involved in mood regulation (prefrontal cortex, hippocampus, striatum), and therefore many interventions that prevent ketamine's "mania-like" side-effects are thought to have potent antioxidant or neuroprotective properties.

For example, a series of studies from Gazal and colleagues explored dietary antioxidants as prophylactics in a ketamine mania model. In multiple studies, pre-treatment with lithium or a blueberry extract markedly protected rats against ketamine-induced hyperlocomotive and oxidative/inflammatory effects in the cortex, striatum, and hippocampus [26,27]. Additionally, pre-treatment with an extract of blackberry (another anthocyanin-rich berry) attenuated ketamine-induced hyperlocomotion and brain oxidative stress markers in a manner similar to lithium [28]. Pretreatment with curcumin [29] (the active component of turmeric), an extract of *Cecropia pachystachya* [30], or gallic acid [31], has also been shown to limit ketamine's hyperlocomotive effects and oxidative/inflammatory markers, particularly in the cortex, hippocampus, and striatum, to a degree similar to pre-treatment with lithium. Gallic acid was also able to normalize acetylcholinesterase activity in the hippocampus and striatum, countering ketamine's increase in acetylcholinesterase activity, as was lithium [31]. However, in another study, the free-radical ederavone, unlike lithium, failed to prevent hyperlocomotion in the open-field test [32]. Taken together, these findings suggest that lithium (and other antioxidants) limit ketamine's hyperlocomotive effects (which, in this model, represent mania), and may attenuate the oxidative/inflammatory responses of ketamine administration.

Beyond oxidative/inflammatory pathways, lithium may limit ketamine-induced hyperlocomotion through other mechanisms. For example, lithium has been shown to limit widespread ketamine-induced forebrain activation in limbic regions, such as the hippocampus, amygdala, lateral septum, and hypothalamus [33], suggesting that lithium can dampen the aberrant circuit excitability underlying ketamine's "mania-like" effect, potentially explaining lithium's clinical anti-manic efficacy. To some degree, lithium-related attenuation of ketamine-induced hyperlocomotion may be related to lithium downregulating PI3K–Akt signaling in the medial PFC [34]. That same study found that pharmacological inhibition of mTOR (with rapamycin) had no effect on hyperlocomotion [34]. It is therefore possible that lithium counteracts ketamine's hyperlocomotive effects, not only via the antioxidant mechanisms mentioned in the previous section, but also via the dampening of neural excitability, potentially via activity in the PI3K–Akt pathway.

To better understand the mechanistic convergence and divergence between ketamine and lithium, it may also be helpful to examine the mechanisms of alternative mood stabilizers and antipsychotics when co-administered with ketamine. For example, lamotrigine has been found to prevent ketamine-induced deficits in prepulse inhibition (PPI), which is an operational measure of sensorimotor gating that is reduced in mania and schizophrenia [35]. Interestingly, in that study, lamotrigine did not reduce the amphetamine-induced PPI deficit, implying that lamotrigine specifically antagonizes glutamate/NMDA-related gating disturbance (i.e., ketamine's PPI effect) but not dopamine-driven PPI deficit (i.e., due to amphetamine). Clinically, this may align with lamotrigine's efficacy in bipolar maintenance, supporting a glutamatergic modulation role in normalizing circuit function. On the other hand, the atypical antipsychotic clozapine may reverse hippocampal-prefrontal impairments in long-term potentiation (LTP) induced by the administration of ketamine [36]. While both ketamine and clozapine increased phospho-GSK3β in the PFC in that study, clozapine uniquely increased GluA1 phosphorylation in the PFC and hippocampus (enhancing AMPA channel conductance at those sites). This pattern indicates an interesting overlap between ketamine, clozapine, and lithium with respect to the effects on GSK3β, suggesting that clozapine and lamotrigine's effects on countering ketamine-related mania/psychosis may be mediated by the effects on glutamatergic transmission.

Using EEG, Bowman et al. [37] used freely moving rats to examine how clozapine and another modulator, naltrexone, influenced ketamine-induced network oscillations, with the clinically important implication being that ketamine's EEG signature of psychosis vs. antidepressant response might be separable. Ketamine alone produced distinct resting-state EEG abnormalities, including a reduction in low-beta power and an increase in gamma- and high-frequency oscillations during resting states (identified via head-mounted accelerometers). Clozapine reversed the low-beta deficit, which may be related to its antipsychotic action. Clozapine also potentiated ketamine's high-frequency oscillation increases, whereas naltrexone (which in human patients blocks ketamine's antidepressant and anti-suicidal effects) [38] specifically blunted the high-frequency oscillation component, which may thus be a marker of ketamine's antidepressant action.

Taken together, these preclinical studies suggest lithium is effective at augmenting the antidepressant effect of ketamine while also proving effective at curtailing ketamine's inducing of mania-like states. This creates a complex preclinical picture, leaving the relationship between acute ketamine response and long-term lithium prophylaxis an open clinical question.

3.2. Clinical Studies

3.2.1. Rapid Antidepressant Efficacy of Ketamine in Mood Disorders

Ketamine has demonstrated significant antidepressant effects in both treatment-resistant bipolar and unipolar depression. In the first randomized placebo-controlled trial in bipolar depression, a single sub-anesthetic IV dose of ketamine (0.5 mg/kg) administered to patients receiving stable levels of lithium or valproate produced significantly improved depressive symptoms within 40 min, with 71% of the 17 patients responding at two weeks (compared to only 6% of the 16 patients in the placebo arm) [39]. There was no treatment response stratification by lithium or valproate treatment status. A replication study confirmed these findings [40], and overall these add-on trials (with patients maintained on lithium or valproate throughout) indicate that ketamine's efficacy is compatible with concurrent mood stabilizer therapy. Furthermore, no relationship has been found between blood levels of lithium or valproate and the antidepressant response to ketamine [41] (although this study may have been underpowered). An open-label clinical series in 53 patients with bipolar depression on stable doses of mood stabilizers further corroborated the concurrent effectiveness of ketamine, with no clear stratification of response across bipolar subtypes and illness histories, number of prior episodes, or concomitant lithium/quetiapine use [42]. Still another study found that lithium or anticonvulsant use was unrelated to response trajectories in patients with severe depression (in a mixed sample of unipolar and bipolar depression) [43]. Taken together, these studies suggest that the acute antidepressant effects of ketamine are not affected by mood stabilizer use.

3.2.2. Symptom Subtypes and Clinical Predictors of Response

Interestingly, ketamine's antidepressant efficacy may differ depending on the depression subtype. In an observational study of 97 patients with treatment-resistant depression, Wang et al. [14] found that those with melancholic features were significantly less likely to respond or to achieve remission, and improved more slowly compared to patients with anxious or non-melancholic depression following six ketamine infusions. These findings suggest that patients with melancholic features might require additional measures or may respond less robustly to ketamine's antidepressant mechanism, whereas those with significant anxious distress (at least in the absence of melancholic signs) could be particularly good candidates for ketamine therapy. While there is difficulty in clinical settings in distinguishing anxious distress from mixed features in depression [44], this finding remains of particular interest given that prior studies have suggested that melancholic features may predict favorable outcomes with lithium prophylaxis [45], indicating further clinical divergence between predictors of lithium and ketamine response. Similarly, in bipolar depression, anxiety comorbidity does not appear to diminish ketamine's efficacy. A post hoc analysis focusing on bipolar patients with high baseline anxiety found that patients with anxious bipolar depression ($n = 21$) derived a similar magnitude and timing of benefit from ketamine as those with non-anxious ($n = 15$) bipolar depression [15]. This efficacy of ketamine in anxious bipolar depression indicates it may bypass some of the limitations of conventional treatments in these populations.

Beyond mood subtype, other clinical factors like personality and trauma history have emerged as potential moderators of ketamine response. A recent study of intranasal ketamine (esketamine) in depressed inpatients reported that patients with a co-morbid personality disorder (the most common of which was borderline personality disorder) had a significantly lower response rate to ketamine, whereas patients with a history of childhood trauma had a higher likelihood of responding [46]. Although this study was exploratory in nature and prone to the risk of uncorrected multiple comparisons, and inclusive of a mixed unipolar/bipolar sample, it poses an interesting further potential contrast between the clinical predictors of response to ketamine and prophylactic lithium therapy, as histories of childhood trauma are predictive of a poorer prophylactic response to lithium [47]. Given that there is some evidence that childhood maltreatment may not decrease responsiveness to anticonvulsant mood stabilizers, it is of potential interest to consider whether acute ketamine antidepressant response may be differentially predictive of response to prophylactic anticonvulsant therapy [48].

An analysis by Pennybaker et al. [49] examined which clinical features distinguished longer-lasting (what they called "extended") antidepressant responses to ketamine, finding that extended responders were both more likely to have a family history of alcohol use disorder in first-degree relatives and to experience greater acute dissociation during the ketamine infusion [49]. Another study found that ketamine responders may have a higher intra-infusion increase in systolic blood pressure compared to non-responders, although the effect on the duration of response was not reported [50]. No clinical predictors that are known to influence prophylactic lithium response in bipolar disorder were identified. Thus, it remains unclear whether the duration of ketamine's antidepressant response provides any information about the likelihood of a subsequent prophylactic lithium response. There has also been mixed evidence of clinical laboratory measures, such as vitamin B12 and VEGF predicting response to ketamine [51,52], with no known overlap with predictors of lithium response.

3.2.3. Adjunctive Treatments to Sustain Ketamine's Effects

Because ketamine's antidepressant benefit is often transient, adjunctive strategies to prolong its effect have been tested. To date, the only randomized controlled trial investigating lithium continuation following ketamine response in patients with treatment-resistant unipolar depression [53] found that lithium offered no advantage over the placebo in maintaining antidepressant response. This suggests that lithium alone, started immediately after ketamine, does not reliably prevent early relapse. Notably, the mean plasma level of lithium in the study was 0.61 mEq/L, toward the lower end of target ranges for both the acute treatment of bipolar depression and the maintenance treatment typically recommended in the guidelines [1]. By contrast, in a 5-year observational follow-up study in 16 patients, Amiaz et al. [54] found that patients receiving quetiapine experienced a longer time until relapse compared to patients receiving other neuroleptics. Patients (a mixed sample of unipolar and bipolar disorder) who received ketamine while on quetiapine had a median time to relapse of ~966 days compared to ~80 days for those on other antipsychotics (aripiprazole, olanzapine, ziprasidone, perphenazine, and clozapine).

3.2.4. Neuroimaging and Neurophysiological Mechanisms

Neuroimaging studies have suggested that ketamine's antidepressant response in BD may be related to the rapid reorganization of brain networks. In a PET imaging study of depressed bipolar patients who did not respond to 4-week trials of either lithium or valproate, and who subsequently underwent a double-blind crossover trial of ketamine vs. saline placebo, Nugent et al. [55] showed that antidepressant responders exhibited the greatest post-infusion increases in ventral striatal metabolism, while higher activity in the subgenual anterior cingulate cortex (sgACC) following the placebo infusion predicted greater symptom relief following ketamine. Along these lines, resting-state fMRI findings in patients with MDD have found that ketamine responders showed lower baseline functional connectivity between the lateral prefrontal cortex and the sgACC, as well as larger increases following treatment. Indeed, Chen et al. [56] also found that patients with treatment-resistant depression showed lower fronto-striatal functional connectivity, and that the magnitude of this connectivity was inversely related to the degree of antidepressant improvement following IV ketamine [56]. Furthermore, in a repeated-infusion protocol, patients with unipolar depression who responded to IV ketamine demonstrated elevated baseline resting-state functional connectivity between various regions and the amygdala [57]. These differences may be related to oscillatory patterns of activity in depression, since Cao et al. [58] reported that responders to IV ketamine showed lower frontal power in the theta band at baseline, but higher treatment-related increases in alpha power, as well as lower alpha asymmetry and theta cordance. The degree to which these findings are associated with a prophylactic response to lithium is unclear, as we are unaware of evidence associating functional connectivity and EEG-biomarkers with prophylactic mood stabilizer response.

3.2.5. Peripheral Biomarkers and Molecular Pathways

One striking finding involves ketamine's effect on the glutamate–mTOR signaling pathway, which is involved in synaptic plasticity, and may be a significant mediator of ketamine's antidepressant action. In a small study of 27 depressed women, Berner et al. [59] found that higher baseline levels of phosphorylated p70S6 kinase (p70S6K, a downstream effector of mTORC1 whose phosphorylation levels served as a proxy for MTORC1 activation in monocytes) were correlated with better antidepressant outcomes from ketamine, although the study used a machine learning method that in such a small sample would be prone to overfitting. Notwithstanding this limitation, the implication here may be that individuals with higher baseline MTORC1 activity may be potentially more "primed" to respond to the antidepressant effects of ketamine, which involves synaptogenesis and may depend on mTOR-mediated pathways. mTOR may also serve as a convergence point between the acute antidepressant effects of ketamine and the longer-term prophylactic effects of lithium in bipolar disorder, since a key mechanism of the latter involves GSK3β inhibition, an upstream regulator of mTOR activity.

In a placebo-controlled clinical study, a single ketamine infusion significantly increased vascular endothelial growth factor A (*VEGFA*) gene expression in whole blood of depressed patients compared to a midazolam (sedative) control [60]. Ketamine also selectively increased the ratio of *VEGFA* to its anti-angiogenic counterpart *PEDF* (pigment epithelial-derived factor, with gene name *SERPINF1*) in peripheral blood, measured about 4 h post-infusion [60]. The specificity of these biomarker changes in the ketamine infusion group may suggest that ketamine engages pro-angiogenic growth factor signaling. Interestingly, in rats, it has been demonstrated that lithium promotes VEGF expression in brain endothelium and astrocytes [61].

In a study by Permoda-Osip et al. [52] of 42 depressed patients, personal/family history of alcohol abuse, elevated serum vitamin B12 concentrations, and elevated serum vascular endothelial growth factor predicted response to ketamine. Villaseñor et al. [62] found that lysophosphatidylethanolamines and lysophosphatidylcholines were increased in ketamine responders relative to non-responders in a sample of patients with bipolar depression, suggesting that mitochondrial fatty acid metabolism may predict response to ketamine. Słupski et al. [63] showed that repeated ketamine infusions decreased serum copper concentrations in depressed patients, though no clear connection to treatment response was found. Reduced copper concentrations may be due to a reduction in the acute phase response, further suggesting ketamine exerts anti-inflammatory effects.

4. Discussion

This review examined the relationship between the acute antidepressant effects of ketamine and the long-term prophylactic outcomes with lithium in bipolar disorder (BD). We found that, while ketamine and lithium share convergent molecular targets that enhance structural plasticity, their clinical predictors of efficacy largely diverge. In this section, we synthesize this evidence, attempt to consider why such divergence may exist, and discuss the implications for clinical prediction and the design of future studies leveraging ketamine response to predict prophylactic response to lithium.

Ketamine and lithium act on overlapping molecular pathways implicated in neuroplasticity. Ketamine produces rapid antidepressant effects through NMDA receptor antagonism, resulting in glutamate surges and AMPA receptor activation, and ultimately downstream mTORC1–BDNF signaling, which together promote rapid synaptogenesis [64,65]. Over a longer time scale, lithium exerts pro-synaptic effects via GSK-3β inhibition, disinhibiting mTOR and supporting neurotrophic and gene-expression changes that sustain long-term mood stabilization [19,66]. Rodent studies have suggested that these mechanisms may compound each other's effects. Indeed, lithium potentiates and prolongs the antidepressant effects of subthreshold ketamine doses, and reduces ketamine-induced oxidative stress [19–21]. Both agents upregulate phosphorylation of GSK-3β and mTOR in the prefrontal cortex, reinforcing structural plasticity [19]. As a result, lithium attenuates ketamine-induced hyperlocomotion and limbic hyperactivation, potentially consistent with its anti-manic effects in humans [33,34]. These findings support a shared capacity to enhance synaptic strength and, especially for lithium, protect against excitotoxicity, suggesting some mechanistic overlap among biological responders. In terms of clinical overlap in function, it is notable that both agents reduce suicidal ideation, albeit over different time scales (ketamine within hours [40,67] and lithium across years of maintenance [68]). A corollary question that arises from this overlap therefore concerns the degree to which synaptic plasticity is involved in the pathophysiology of suicidal ideation and behavior.

Despite the aforementioned clinical overlaps, the clinical phenotypes of ketamine and lithium responders differ significantly. Excellent lithium responders typically present with "classical" BD, characterized by episodic euphoric manias, discrete illness episodes, complete inter-episode remission, later onset, and strong family history [69,70]. This stands in stark contrast to the atypical clinical phenotypes that are ostensibly characteristic of ketamine responders and lithium non-response. Indeed, while high body mass index (BMI) predicts ketamine response [17,71], factors such as insulin resistance and metabolic syndrome are associated with lithium non-response [10,72]. Similarly, while comorbid anxiety and mixed features limit lithium prophylaxis [4,6,9], they are associated with positive ketamine response [15,16]. Childhood trauma predicts poor lithium outcomes [47] but may enhance ketamine responsiveness [46]. Finally, depression subtypes diverge: melancholic features favor lithium [45], whereas non-melancholic or anxious depressions respond preferentially to ketamine [14]. Taken together, these findings suggest that ketamine responders are overrepresented with respect to the features that portend worse prognosis for longer-term prophylactic lithium responsiveness.

Ultimately, our review failed to identify clear preclinical evidence explaining the divergent phenotypic profiles of responders to acute ketamine treatment and long-term lithium prophylaxis. Animal models reliably show that ketamine produces rapid acute antidepressant effects which can be extended or modulated by lithium, but do not provide clinically relevant approaches to identifying whether and which ketamine responders are less likely to benefit from lithium prophylaxis. For example, while oxidative stress and inflammatory changes accompany ketamine's behavioral effects in rodents [26–28], these findings do not straightforwardly map onto the clinical observation that ketamine responders often show the phenotypic predictors of lithium non-responders. Similarly, while insulin and metabolic pathways are implicated in both ketamine's and lithium's actions [22], animal studies do not yet clarify why metabolic dysregulation (e.g., high BMI) would favor acute ketamine response but undermine lithium prophylaxis [10].

One possible explanation for our failure to reconcile the preclinical and clinical data is that preclinical models largely capture the acute effects of ketamine and lithium, such as how lithium extends the duration of the antidepressant effect of ketamine or counteracts ketamine-induced hyperlocomotion. By contrast, the clinical divergence in response emerges over years of illness course (since lithium responsiveness requires more than 1–2 years to elucidate), reflecting potential differences in the mechanisms of long-term disease stabilization versus short-term symptom relief. Unfortunately, animal models are not well suited to capturing these long-horizon prophylactic outcomes. This limits our ability to test hypotheses about divergent predictors of response at the biological level. Thus, while the clinical data suggest an inverse predictive relationship between ketamine response and prophylactic lithium response, the preclinical literature remains largely neutral on this matter.

In summary, convergent mechanisms of action between ketamine and lithium in BD include action on GSK-3β/mTOR pathways as well as the enhancement of synaptic plasticity. Ketamine-induced oxidative stress and hyperlocomotion are consistently limited by lithium. Clinical predictors of response seem to diverge; ketamine response is associated with metabolic risk factors, anxiety/mixed features, and non-melancholic presentations, which are generally predictors of prophylactic lithium non-response. Taken together, this suggests the potential for a personalized medicine approach, whereby response to ketamine in patients with bipolar depression could serve as a biomarker of non-response to lithium (pending prospective validation), saving individuals prolonged ineffective trials of lithium prophylaxis.

Several limitations must be acknowledged. No clinical study has directly tested ketamine response as a predictor of lithium prophylaxis, so all inferences remain indirect. Additionally, many of the clinical studies included herein included mixed unipolar/bipolar samples, which were not disentangled in post hoc analyses. Third, and arguably most significantly, ketamine trials are likely enriched for treatment-resistant patients, implying that they may contain a higher proportion of lithium non-responders, which may inflate the appearance of divergence. That is, due to ketamine typically being reserved for treatment-resistant populations, many patients receiving ketamine are a priori lithium non-responders due to previously failed lithium trials. As prior medication trials were scarcely defined in the clinical studies used within this evidence synthesis, it is unclear how many of the study populations examined were enriched for lithium non-responders. Additionally, this temporal relationship of medication usage is incongruent with the clinical question initially outlined—that is, is responsiveness to ketamine predictive of responsiveness to lithium.

What may appear to be a divergence in the clinical phenotype of responders to ketamine vs. lithium may be an artifact of sampling bias, whereby samples enriched for lithium non-responsiveness who respond to ketamine (or any other treatment) will appear to diverge from classical lithium responders with respect to clinical, neurophysiological, and biochemical factors. Furthermore, many predictor studies, both for ketamine and lithium, are small, underpowered, and heterogeneous in design, complicating interpretation. Finally, one must also note that preclinical studies, while mechanistically valuable, cannot model the full complexity of BD or the longitudinal trajectories that are required to understand prophylactic lithium response, limiting their utility in understanding this critical element of disease management in humans.

5. Conclusions

The existing literature supports mechanistic overlap but clinical divergence between ketamine and lithium responders. No studies directly examined ketamine response as a predictor of lithium nonresponse, and using existing data to investigate this hypothesis comes with many limitations. Most notably, ketamine trials are enriched for treatment-resistant populations, and clinical divergence between ketamine and lithium responders may simply be reflective of ketamine responders being a priori lithium non-responders. To resolve these uncertainties, prospective, longitudinal studies must be conducted in which individuals with BD who have received acute courses of ketamine for depression are followed prospectively while receiving lithium prophylaxis. Incorporating biomarkers such as inflammatory and metabolic measures, neuroimaging, and electrophysiology may help link clinical observations to mechanistic pathways, and induced pluripotent stem cell-based assays may help qualify overlapping and diverging mechanistic pathways at the individual level [73–77]. Preclinical work should additionally focus on developing animal models of the longitudinal relapsing and remitting trajectory of BD, to better understand prophylaxis. Ultimately, such studies would serve to determine whether ketamine response, which could be a rapid and easily identified biomarker, predicts response to lithium prophylaxis, thereby potentially saving patients years of ineffective trials of prophylactic therapy. If validated in prospective cohorts, such a precision medicine approach could help establish control over the disease earlier, improving both morbidity and mortality in this patient population.

Author Contributions: J.T. and A.N. conducted the data collection and screening, as well as manuscript preparation. J.E.A.C. assisted with manuscript edits. All authors have read and agreed to the published version of the manuscript.

Funding: This research was funded by the QEII Foundation Translating Research into Care (TRIC) Grant, the Nova Scotia Health Innovation Catalyst Fund, and the Nova Scotia Health Research Fund.

Data Availability Statement: No new data were created or analyzed in this study. Data sharing is not applicable.

Conflicts of Interest: The authors declare no conflict of interest.

References

1. Yatham, L.N.; Kennedy, S.H.; Parikh, S.V.; Schaffer, A.; Bond, D.J.; Frey, B.N.; Sharma, V.; Goldstein, B.I.; Rej, S.; Beaulieu, S.; et al. Canadian Network for Mood and Anxiety Treatments (CANMAT) and International Society for Bipolar Disorders (ISBD) 2018 guidelines for the management of patients with bipolar disorder. *Bipolar Disord.* **2018**, *20*, 97–170. [CrossRef]
2. Grof, P. Sixty Years of Lithium Responders. *Neuropsychobiology* **2010**, *62*, 8–16. [CrossRef]
3. Garnham, J.; Munro, A.; Slaney, C.; MacDougall, M.; Passmore, M.; Duffy, A.; O'DOnovan, C.; Teehan, A.; Alda, M. Prophylactic treatment response in bipolar disorder: Results of a naturalistic observation study. *J. Affect. Disord.* **2007**, *104*, 185–190. [CrossRef]

4. Nunes, A.; Ardau, R.; Berghöfer, A.; Bocchetta, A.; Chillotti, C.; Deiana, V.; Garnham, J.; Grof, E.; Hajek, T.; Manchia, M.; et al. Prediction of lithium response using clinical data. *Acta Psychiatr. Scand.* **2020**, *141*, 131–141. [CrossRef]
5. Scott, K.; Khayachi, A.; Alda, M.; Nunes, A. Prediction of Treatment Outcome in Bipolar Disorder: When Can We Expect Clinical Relevance? *Biol. Psychiatry* **2025**, *98*, 285–292. [CrossRef] [PubMed]
6. Nunes, A.; Stone, W.; Ardau, R.; Berghöfer, A.; Bocchetta, A.; Chillotti, C.; Deiana, V.; Degenhardt, F.; Forstner, A.J.; Garnham, J.S.; et al. Exemplar scoring identifies genetically separable phenotypes of lithium responsive bipolar disorder. *Transl. Psychiatry* **2021**, *11*, 36. [CrossRef] [PubMed]
7. Gershon, S.; Chengappa, K.R.; Malhi, G.S. Lithium specificity in bipolar illness: A classic agent for the classic disorder. *Bipolar Disord.* **2009**, *11* (Suppl. 2), 34–44. [CrossRef]
8. Hui, T.P.; Kandola, A.; Shen, L.; Lewis, G.; Osborn, D.P.J.; Geddes, J.R.; Hayes, J.F. A systematic review and meta-analysis of clinical predictors of lithium response in bipolar disorder. *Acta Psychiatr. Scand.* **2019**, *140*, 94–115. [CrossRef]
9. Lin, Y.; Maihofer, A.X.; Stapp, E.; Ritchey, M.; Alliey-Rodriguez, N.; Anand, A.; Balaraman, Y.; Berrettini, W.H.; Bertram, H.; Bhattacharjee, A.; et al. Clinical predictors of non-response to lithium treatment in the Pharmacogenomics of Bipolar Disorder (PGBD) study. *Bipolar Disord.* **2021**, *23*, 821–831. [CrossRef]
10. Calkin, C.V.; Ruzickova, M.; Uher, R.; Hajek, T.; Slaney, C.M.; Garnham, J.S.; O'Donovan, M.C.; Alda, M. Insulin resistance and outcome in bipolardisorder. *Br. J. Psychiatry* **2015**, *206*, 52–57. [CrossRef]
11. Fancy, F.; Haikazian, S.; Johnson, D.E.; Chen-Li, D.C.J.; Levinta, A.; Husain, M.I.; Mansur, R.B.; Rosenblat, J.D. Ketamine for bipolar depression: An updated systematic review. *Ther. Adv. Psychopharmacol.* **2023**, *13*, 20451253231202723. [CrossRef]
12. Kraus, C.; Rabl, U.; Vanicek, T.; Carlberg, L.; Popovic, A.; Spies, M.; Bartova, L.; Gryglewski, G.; Papageorgiou, K.; Lanzenberger, R.; et al. Administration of ketamine for unipolar and bipolar depression. *Int. J. Psychiatry Clin. Pract.* **2017**, *21*, 2–12. [CrossRef] [PubMed]
13. Lam, R.W. Onset, time course and trajectories of improvement with antidepressants. *Eur. Neuropsychopharmacol.* **2012**, *22*, S492–S498. [CrossRef] [PubMed]
14. Wang, C.; Zhou, Y.; Zheng, W.; Liu, W.; Zhan, Y.; Li, H.; Chen, L.; Zhang, B.; Walter, M.; Li, M.; et al. Association between depression subtypes and response to repeated-dose intravenous ketamine. *Acta Psychiatr. Scand.* **2019**, *140*, 446–457. [CrossRef] [PubMed]
15. Ionescu, D.F.; Luckenbaugh, D.A.; Niciu, M.J.; Richards, E.M.; Zarate, C.A., Jr. A single infusion of ketamine improves depression scores in patients with anxious bipolar depression. *Bipolar Disord.* **2015**, *17*, 438–443. [CrossRef]
16. McIntyre, R.S.; Lipsitz, O.; Rodrigues, N.B.; Lee, Y.; Cha, D.S.; Vinberg, M.; Lin, K.; Malhi, G.S.; Subramaniapillai, M.; Kratiuk, K.; et al. The effectiveness of ketamine on anxiety, irritability, and agitation: Implications for treating mixed features in adults with major depressive or bipolar disorder. *Bipolar Disord.* **2020**, *22*, 831–840. [CrossRef]
17. Rong, C.; Park, C.; Rosenblat, J.D.; Subramaniapillai, M.; Zuckerman, H.; Fus, D.; Lee, Y.L.; Pan, Z.; Brietzke, E.; Mansur, R.B.; et al. Predictors of Response to Ketamine in Treatment Resistant Major Depressive Disorder and Bipolar Disorder. *Int. J. Environ. Res. Public Health* **2018**, *15*, 771. [CrossRef]
18. Tricco, A.C.; Lillie, E.; Zarin, W.; O'Brien, K.K.; Colquhoun, H.; Levac, D.; Moher, D.; Peters, M.D.J.; Horsley, T.; Weeks, L.; et al. PRISMA Extension for Scoping Reviews (PRISMA-ScR): Checklist and Explanation. *Ann. Intern. Med.* **2018**, *169*, 467–473. [CrossRef]
19. Chiu, C.T.; Scheuing, L.; Liu, G.; Liao, H.M.; Linares, G.R.; Lin, D.; Chuang, D.-M. The Mood Stabilizer Lithium Potentiates the Antidepressant-Like Effects and Ameliorates Oxidative Stress Induced by Acute Ketamine in a Mouse Model of Stress. *Int. J. Neuropsychopharmacol.* **2015**, *18*, pyu102. [CrossRef]
20. Liu, R.J.; Fuchikami, M.; Dwyer, J.M.; Lepack, A.E.; Duman, R.S.; Aghajanian, G.K. GSK-3 Inhibition Potentiates the Synaptogenic and Antidepressant-Like Effects of Subthreshold Doses of Ketamine. *Neuropsychopharmacology* **2013**, *38*, 2268–2277. [CrossRef]
21. do Vale, E.M.; Xavier, C.C.; Nogueira, B.G.; Campos, B.C.; de Aquino, P.E.A.; da Costa, R.O.; Leal, L.K.A.M.; de Vasconcelos, S.M.M.; Neves, K.R.T.; Viana, G.S.d.B. Antinociceptive and Anti-Inflammatory Effects of Ketamine and the Relationship to Its Antidepressant Action and GSK3 Inhibition. *Basic Clin. Pharmacol. Toxicol.* **2016**, *119*, 562–573. [CrossRef]
22. Price, J.B.; Yates, C.G.; Morath, B.A.; Van De Wakker, S.K.; Yates, N.J.; Butters, K.; Frye, M.A.; McGee, S.L.; Tye, S.J. Lithium augmentation of ketamine increases insulin signaling and antidepressant-like active stress coping in a rodent model of treatment-resistant depression. *Transl. Psychiatry* **2021**, *11*, 598. [CrossRef] [PubMed]
23. Maltbie, E.; Gopinath, K.; Urushino, N.; Kempf, D.; Howell, L. Ketamine-induced brain activation in awake female nonhuman primates: A translational functional imaging model. *Psychopharmacology* **2016**, *233*, 961–972. [CrossRef] [PubMed]
24. Mayberg, H.S.; Lozano, A.M.; Voon, V.; McNeely, H.E.; Seminowicz, D.; Hamani, C.; Schwalb, J.M.; Kennedy, S.H. Deep Brain Stimulation for Treatment-Resistant Depression. *Neuron* **2005**, *45*, 651–660. [CrossRef] [PubMed]
25. Stepan, J.; Hladky, F.; Uribe, A.; Holsboer, F.; Schmidt, M.V.; Eder, M. High-Speed imaging reveals opposing effects of chronic stress and antidepressants on neuronal activity propagation through the hippocampal trisynaptic circuit. *Front. Neural Circuits* **2015**, *9*, 70. [CrossRef]

26. Debom, G.; Gazal, M.; Soares, M.S.P.; do Couto, C.A.T.; Mattos, B.; Lencina, C.; Kaster, M.P.; Ghisleni, G.C.; Tavares, R.; Braganhol, E.; et al. Preventive effects of blueberry extract on behavioral and biochemical dysfunctions in rats submitted to a model of manic behavior induced by ketamine. *Brain Res. Bull.* **2016**, *127*, 260–269. [CrossRef]
27. Spohr, L.; Soares, M.S.P.; Oliveira, P.S.; da Silveira de Mattos, B.; Bona, N.P.; Pedra, N.S.; Teixeira, F.C.; Couto, C.A.T.D.; Chaves, V.C.; Reginatto, F.H.; et al. Combined actions of blueberry extract and lithium on neurochemical changes observed in an experimental model of mania: Exploiting possible synergistic effects. *Metab. Brain Dis.* **2019**, *34*, 605–619. [CrossRef]
28. Chaves, V.C.; Soares, M.S.P.; Spohr, L.; Teixeira, F.; Vieira, A.; Constantino, L.S.; Pizzol, F.D.; Lencina, C.L.; Spanevello, R.M.; Freitas, M.P.; et al. Blackberry extract improves behavioral and neurochemical dysfunctions in a ketamine-induced rat model of mania. *Neurosci. Lett.* **2020**, *714*, 134566. [CrossRef]
29. Gazal, M.; Valente, M.R.; Acosta, B.A.; Kaufmann, F.N.; Braganhol, E.; Lencina, C.L.; Stefanello, F.M.; Ghisleni, G.; Kaster, M.P. Neuroprotective and antioxidant effects of curcumin in a ketamine-induced model of mania in rats. *Eur. J. Pharmacol.* **2014**, *724*, 132–139. [CrossRef]
30. Gazal, M.; Kaufmann, F.N.; Acosta, B.A.; Oliveira, P.S.; Valente, M.R.; Ortmann, C.F.; Sturbelle, R.; Lencina, C.L.; Stefanello, F.M.; Kaster, M.P.; et al. Preventive Effect of Cecropia pachystachya Against Ketamine-Induced Manic Behavior and Oxidative Stress in Rats. *Neurochem. Res.* **2015**, *40*, 1421–1430. [CrossRef]
31. Recart, V.M.; Spohr, L.; Soares, M.S.P.; de Mattos Bda, S.; Bona, N.P.; Pedra, N.S.; Teixeira, F.C.; Gamaro, G.D.; Stefanello, F.; Spanevello, R. Gallic acid protects cerebral cortex, hippocampus, and striatum against oxidative damage and cholinergic dysfunction in an experimental model of manic-like behavior: Comparison with lithium effects. *Int. J. Dev. Neurosci.* **2021**, *81*, 167–178. [CrossRef]
32. Arslan, F.C.; Tiryaki, A.; Yıldırım, M.; Özkorumak, E.; Alver, A.; Altun, İ.K.; İnce, I.; Gedikli, Ö. The effects of edaravone in ketamine-induced model of mania in rats. *Acta Neurobiol. Exp.* **2016**, *76*, 192–198. [CrossRef] [PubMed]
33. Gao, T.H.; Ni, R.J.; Liu, S.; Tian, Y.; Wei, J.; Zhao, L.; Wang, Q.; Ni, P.; Ma, X.; Li, T. Chronic lithium exposure attenuates ketamine-induced mania-like behavior and c-Fos expression in the forebrain of mice. *Pharmacol. Biochem. Behav.* **2021**, *202*, 173108. [CrossRef] [PubMed]
34. Ni, R.J.; Gao, T.H.; Wang, Y.Y.; Tian, Y.; Wei, J.X.; Zhao, L.S.; Ni, P.-Y.; Ma, X.-H.; Li, T. Chronic lithium treatment ameliorates ketamine-induced mania-like behavior via the PI3K-AKT signaling pathway. *Zool. Res.* **2022**, *43*, 989–1004. [CrossRef] [PubMed]
35. Brody, S.A.; Geyer, M.A.; Large, C.H. Lamotrigine prevents ketamine but not amphetamine-induced deficits in prepulse inhibition in mice. *Psychopharmacology* **2003**, *169*, 240–246. [CrossRef]
36. Rame, M.; Caudal, D.; Schenker, E.; Svenningsson, P.; Spedding, M.; Jay, T.M.; Godsil, B.P. Clozapine counteracts a ketamine-induced depression of hippocampal-prefrontal neuroplasticity and alters signaling pathway phosphorylation. *PLoS ONE* **2017**, *12*, e0177036. [CrossRef]
37. Bowman, C.; Richter, U.; Jones, C.R.; Agerskov, C.; Herrik, K.F. Activity-State Dependent Reversal of Ketamine-Induced Resting State EEG Effects by Clozapine and Naltrexone in the Freely Moving Rat. *Front. Psychiatry* **2022**, *13*, 737295. [CrossRef]
38. Williams, N.R.; Heifets, B.D.; Blasey, C.; Sudheimer, K.; Pannu, J.; Pankow, H.; Hawkins, J.; Birnbaum, J.; Lyons, D.M.; Rodriguez, C.I.; et al. Attenuation of Antidepressant Effects of Ketamine by Opioid Receptor Antagonism. *Am. J. Psychiatry* **2018**, *175*, 1205–1215. [CrossRef]
39. Diazgranados, N.; Ibrahim, L.; Brutsche, N.E.; Newberg, A.; Kronstein, P.; Khalife, S.; Kammerer, W.A.; Quezado, Z.; Luckenbaugh, D.A.; Salvadore, G.; et al. A Randomized Add-on Trial of an N-methyl-d-aspartate Antagonist in Treatment-Resistant Bipolar Depression. *Arch. Gen. Psychiatry* **2010**, *67*, 793–802. [CrossRef]
40. Zarate, C.A.; Brutsche, N.E.; Ibrahim, L.; Franco-Chaves, J.; Diazgranados, N.; Cravchik, A.; Selter, J.; Marquardt, C.A.; Liberty, V.; Luckenbaugh, D.A. Replication of Ketamine's Antidepressant Efficacy in Bipolar Depression: A Randomized Controlled Add-On Trial. *Biol. Psychiatry* **2012**, *71*, 939–946. [CrossRef]
41. Xu, A.J.; Niciu, M.J.; Lundin, N.B.; Luckenbaugh, D.A.; Ionescu, D.F.; Richards, E.M.; Voort, J.L.V.; Ballard, E.D.; Brutsche, N.E.; Machado-Vieira, R.; et al. Lithium and Valproate Levels Do Not Correlate with Ketamine's Antidepressant Efficacy in Treatment-Resistant Bipolar Depression. *Neural Plast.* **2015**, *2015*, 858251. [CrossRef] [PubMed]
42. Rybakowski, J.K.; Permoda-Osip, A.; Bartkowska-Sniatkowska, A. Ketamine augmentation rapidly improves depression scores in inpatients with treatment-resistant bipolar depression. *Int. J. Psychiatry Clin. Pract.* **2017**, *21*, 99–103. [CrossRef] [PubMed]
43. O'Brien, B.; Lijffijt, M.; Lee, J.; Kim, Y.S.; Wells, A.; Murphy, N.; Ramakrishnan, N.; Swann, A.C.; Mathew, S.J. Distinct trajectories of antidepressant response to intravenous ketamine. *J. Affect. Disord.* **2021**, *286*, 320–329. [CrossRef] [PubMed]
44. Tundo, A.; Musetti, L.; Grande, C.D.; de Filippis, R.; Proietti, L.; Marazziti, D.; Gibertoni, D.; Dell'osso, L. The relationship between depression with anxious distress DSM-5 specifier and mixed depression: A network analysis. *CNS Spectr.* **2021**, *26*, 251–257. [CrossRef]
45. Serretti, A.; Lattuada, E.; Franchini, L.; Smeraldi, E. Melancholic features and response to lithium prophylaxis in mood disorders. *Depress. Anxiety* **2000**, *11*, 73–79. [CrossRef]

46. Peters, E.M.; Halpape, K.; Cheveldae, I.; Jacobson, P.; Wanson, A. Predictors of response to intranasal ketamine in patients hospitalized for treatment-resistant depression. *Pers. Med. Psychiatry* **2024**, *43–44*, 100119. [CrossRef]
47. Etain, B.; Lajnef, M.; Brichant-Petitjean, C.; Geoffroy, P.A.; Henry, C.; Gard, S.; Kahn, J.P.; Leboyer, M.; Young, A.H.; Bellivier, F. Childhood trauma and mixed episodes are associated with poor response to lithium in bipolar disorders. *Acta Psychiatr. Scand.* **2017**, *135*, 319–327. [CrossRef]
48. Cascino, G.; D'Agostino, G.; Monteleone, A.M.; Marciello, F.; Caivano, V.; Monteleone, P.; Maj, M. Childhood maltreatment and clinical response to mood stabilizers in patients with bipolar disorder. *Hum. Psychopharmacol. Clin. Exp.* **2021**, *36*, e2783. [CrossRef]
49. Pennybaker, S.J.; Niciu, M.J.; Luckenbaugh, D.A.; Zarate, C.A. Symptomatology and Predictors of Antidepressant Efficacy in Extended Responders to a Single Ketamine Infusion. *J. Affect. Disord.* **2017**, *208*, 560–566. [CrossRef]
50. Keith, K.M.; Geller, J.; Froehlich, A.; Arfken, C.; Oxley, M.; Mischel, N. Vital Sign Changes During Intravenous Ketamine Infusions for Depression: An Exploratory Study of Prognostic Indications. *J. Clin. Psychopharmacol.* **2022**, *42*, 254–259. [CrossRef]
51. Lundin, N.B.; Niciu, M.J.; Luckenbaugh, D.A.; Ionescu, D.F.; Richards, E.M.; Voort, J.L.V.; Brutsche, N.E.; Machado-Vieira, R.; Zarate, C. Baseline Vitamin B12 and Folate Levels Do Not Predict Improvement in Depression After a Single Infusion of Ketamine. *Pharmacopsychiatry* **2014**, *47*, 141–144. [CrossRef]
52. Permoda-Osip, A.; Skibińska, M.; Bartkowska-Sniatkowska, A.; Kliwicki, S.; Chłopocka-Woźniak, M.; Rybakowski, J.K. Factors connected with efficacy of single ketamine infusion in bipolar depression. *Psychiatr. Pol.* **2014**, *48*, 35–47. [CrossRef]
53. Costi, S.; Soleimani, L.; Glasgow, A.; Brallier, J.; Spivack, J.; Schwartz, J.; Levitch, C.F.; Richards, S.; Hoch, M.; Wade, E.; et al. Lithium continuation therapy following ketamine in patients with treatment resistant unipolar depression: A randomized controlled trial. *Neuropsychopharmacology* **2019**, *44*, 1812–1819. [CrossRef] [PubMed]
54. Amiaz, R.; Saporta, R.; Noy, A.; Berkenstadt, H.; Weiser, M. Can Quetiapine Prolong the Antidepressant Effect of Ketamine?: A 5-Year Follow-up Study. *J. Clin. Psychopharmacol.* **2021**, *41*, 673–675. [CrossRef] [PubMed]
55. Nugent, A.C.; Diazgranados, N.; Carlson, P.J.; Ibrahim, L.; Luckenbaugh, D.A.; Brutsche, N.; Herscovitch, P.; Drevets, W.C.; Zarate, C.A., Jr. Neural correlates of rapid antidepressant response to ketamine in bipolar disorder. *Bipolar Disord.* **2014**, *16*, 119–128. [CrossRef] [PubMed]
56. Chen, M.H.; Chang, W.C.; Lin, W.C.; Tu, P.C.; Li, C.T.; Bai, Y.M.; Tsai, S.-J.; Huang, W.-S.; Su, T.-P. Functional Dysconnectivity of Frontal Cortex to Striatum Predicts Ketamine Infusion Response in Treatment-Resistant Depression. *Int. J. Neuropsychopharmacol.* **2020**, *23*, 791–798. [CrossRef]
57. Liu, H.; Wang, C.; Lan, X.; Li, W.; Zhang, F.; Hu, Z.; Ye, Y.; Ning, Y.; Zhou, Y. Functional connectivity of the amygdala subregions and the antidepressant effects of repeated ketamine infusions in major depressive disorder. *Eur. Psychiatry* **2024**, *67*, e33. [CrossRef]
58. Cao, Z.; Lin, C.T.; Ding, W.; Chen, M.H.; Li, C.T.; Su, T.P. Identifying Ketamine Responses in Treatment-Resistant Depression Using a Wearable Forehead EEG. *IEEE Trans. Biomed. Eng.* **2019**, *66*, 1668–1679. [CrossRef]
59. Berner, K.; Oz, N.; Kaya, A.; Acharjee, A.; Berner, J. mTORC1 activation in presumed classical monocytes: Observed correlation with human size variation and neuropsychiatric disease. *Aging* **2024**, *16*, 11134–11150. [CrossRef]
60. McGrory, C.L.; Ryan, K.M.; Gallagher, B.; McLoughlin, D.M. Vascular endothelial growth factor and pigment epithelial-derived factor in the peripheral response to ketamine. *J. Affect. Disord.* **2020**, *273*, 380–383. [CrossRef]
61. Guo, S.; Arai, K.; Stins, M.F.; Chuang, D.M.; Lo, E.H. Lithium Upregulates Vascular Endothelial Growth Factor in Brain Endothelial Cells and Astrocytes. *Stroke* **2009**, *40*, 652–655. [CrossRef]
62. Villaseñor, A.; Ramamoorthy, A.; Silva dos Santos, M.; Lorenzo, M.P.; Laje, G.; Zarate, C., Jr.; Barbas, C.; Wainer, I.W. A pilot study of plasma metabolomic patterns from patients treated with ketamine for bipolar depression: Evidence for a response-related difference in mitochondrial networks. *Br. J. Pharmacol.* **2014**, *171*, 2230–2242. [CrossRef]
63. Słupski, J.; Cubała, W.J.; Górska, N.; Słupska, A.; Gałuszko-Węgielnik, M. Copper Concentrations in Ketamine Therapy for Treatment-Resistant Depression. *Brain Sci.* **2020**, *10*, 971. [CrossRef] [PubMed]
64. Duman, R.S.; Li, N.; Liu, R.J.; Duric, V.; Aghajanian, G. Signaling Pathways Underlying the Rapid Antidepressant Actions of Ketamine. *Neuropharmacology* **2012**, *62*, 35–41. [CrossRef] [PubMed]
65. Zanos, P.; Gould, T.D. Intracellular signaling pathways involved in (S)- and (R)-ketamine antidepressant actions. *Biol. Psychiatry* **2018**, *83*, 2–4. [CrossRef] [PubMed]
66. Quiroz, J.A.; Gould, T.D.; Manji, H.K. Molecular effects of lithium. *Mol. Interv.* **2004**, *4*, 259–272. [CrossRef]
67. Wilkinson, S.T.; Katz, R.B.; Toprak, M.; Webler, R.; Ostroff, R.B.; Sanacora, G. Acute and Longer-Term Outcomes Using Ketamine as a Clinical Treatment at the Yale Psychiatric Hospital. *J. Clin. Psychiatry* **2018**, *79*, 10099. [CrossRef]
68. Cipriani, A.; Hawton, K.; Stockton, S.; Geddes, J.R. Lithium in the prevention of suicide in mood disorders: Updated systematic review and meta-analysis. *BMJ* **2013**, *346*, f3646. [CrossRef]
69. Grof, P.; Duffy, A.; Cavazzoni, P.; Grof, E.; Garnham, J.; MacDougall, M.; O'DOnovan, C.; Alda, M. Is response to prophylactic lithium a familial trait? *J. Clin. Psychiatry* **2002**, *63*, 942–947. [CrossRef]

70. Manchia, M.; Adli, M.; Akula, N.; Ardau, R.; Aubry, J.M.; Backlund, L.; Banzato, C.E.; Baune, B.T.; Bellivier, F.; Bengesser, S.; et al. Assessment of Response to Lithium Maintenance Treatment in Bipolar Disorder: A Consortium on Lithium Genetics (ConLiGen) Report. *PLoS ONE* **2013**, *8*, e65636. [CrossRef]
71. McIntyre, R.S.; Rosenblat, J.D.; Nemeroff, C.B.; Sanacora, G.; Murrough, J.W.; Berk, M.; Brietzke, E.; Dodd, S.; Gorwood, P.; Ho, R.; et al. Synthesizing the Evidence for Ketamine and Esketamine in Treatment-Resistant Depression: An International Expert Opinion on the Available Evidence and Implementation. *Am. J. Psychiatry* **2021**, *178*, 383–399. [CrossRef] [PubMed]
72. Rybakowski, J.K.; Suwalska, A. Excellent lithium responders have normal cognitive functions and plasma BDNF levels. *Int. J. Neuropsychopharmacol.* **2010**, *13*, 617–622. [CrossRef]
73. Khayachi, A.; Ase, A.; Liao, C.; Kamesh, A.; Kuhlmann, N.; Schorova, L.; Chaumette, B.; Dion, P.; Alda, M.; Séguéla, P.; et al. Chronic lithium treatment alters the excitatory/inhibitory balance of synaptic networks and reduces mGluR5–PKC signalling in mouse cortical neurons. *J. Psychiatry Neurosci.* **2021**, *46*, E402–E414. [CrossRef]
74. Khayachi, A.; Abuzgaya, M.; Liu, Y.; Jiao, C.; Dejgaard, K.; Schorova, L.; Kamesh, A.; He, Q.; Cousineau, Y.; Pietrantonio, A.; et al. Akt and AMPK activators rescue hyperexcitability in neurons from patients with bipolar disorder. *EBioMedicine* **2024**, *104*, 105161. [CrossRef]
75. Stern, S.; Santos, R.; Marchetto, M.C.; Mendes, A.P.D.; Rouleau, G.; Biesmans, S.; Wang, Q.-W.; Yao, J.; Charnay, P.; Bang, A.G.; et al. Neurons derived from patients with bipolar disorder divide into intrinsically different sub-populations of neurons, predicting the patients' responsiveness to lithium. *Mol. Psychiatry* **2018**, *23*, 1453–1465. [CrossRef]
76. Stern, S.; Sarkar, A.; Galor, D.; Stern, T.; Mei, A.; Stern, Y.; Mendes, A.P.; Randolph-Moore, L.; Rouleau, G.; Bang, A.G.; et al. A Physiological Instability Displayed in Hippocampal Neurons Derived from Lithium-Nonresponsive Bipolar Disorder Patients. *Biol. Psychiatry* **2020**, *88*, 150–158. [CrossRef]
77. Stern, S.; Sarkar, A.; Stern, T.; Mei, A.; Mendes, A.P.D.; Stern, Y.; Goldberg, G.; Galor, D.; Nguyen, T.; Randolph-Moore, L.; et al. Mechanisms Underlying the Hyperexcitability of CA3 and Dentate Gyrus Hippocampal Neurons Derived from Patients with Bipolar Disorder. *Biol. Psychiatry* **2020**, *88*, 139–149. [CrossRef]

pharmaceuticals

Review

History of Suicide Prevention with Lithium Treatment

Leonardo Tondo [1,2,3,4,*] and Ross J. Baldessarini [1,2]

1 Department of Psychiatry, Harvard Medical School, 116 Mill Street, Boston, MA 02478, USA; rbaldessarini@mclean.harvard.edu
2 International Consortium for Mood & Psychotic Disorder Research, McLean Hospital, Belmont, MA 02478, USA
3 Lucio Bini Mood Disorder Centers, 42 Via Crescenzio, 00,193 Rome, Italy
4 Lucio Bini Mood Disorder Centers, 28 Via Cavalcanti, 09,128 Cagliari, Italy
* Correspondence: ltondo@aol.com; Tel.: +39-335453180

Abstract: Suicidal behavior is prevalent among individuals with psychiatric illnesses, especially mood, substance abuse, and psychotic disorders. Over the past several decades, lithium treatment in patients with mood disorders has been increasingly used to lower the risk of suicidal behavior. This overview considers that lithium treatment has the most abundant evidence of reducing suicidal behavior. It also examines the hypothesis that higher natural lithium levels in drinking water correlate with reduced suicide rates. We report findings from trials comparing lithium treatment with its absence, placebos, or alternative treatments for suicide prevention and address substantial challenges in such studies. The mechanisms behind lithium's potentially protective effects against suicidal behavior remain uncertain. However, it is believed that lithium may produce anti-aggressive/anti-impulsive effects that directly contribute to anti-suicidal outcomes and mood-stabilizing effects that indirectly lead to the same results. Anti-aggressive/anti-impulsive effects may be obtained at the very low levels of lithium present in drinking water, whereas recurrence prevention may be attained at therapeutic levels.

Keywords: attempt; bipolar disorders; lithium; major depressive disorder; suicide

1. Brief History of Lithium-Associated Suicide Prevention

This review aims to describe the various stages that led to demonstrating the efficacy of long-term treatments with lithium salts in preventing suicidal behaviors.

Goodwin and Jamison's book *Manic-Depressive Illness* [1] reported an average risk of 19% (range: 9.0–60%) for lifetime risk of suicide-associated fatalities in patients with bipolar disorder across treatments. It is also well established that bipolar disorder and recurrent major depression have the highest standardized Mortality ratios (SMRs) for suicide out of all psychiatric disorder, averaging approximately 20 times above the international general population suicide rate of about 15/100,000/year (15/100k PEY; 0.015%/year) [2]. Such high rates have been confirmed by other studies [3,4].

Suicidal behavior is strongly associated with the depressive phase of bipolar disorder, especially when mixed (manic–depressive) features are present, so it seems plausible to suggest that antidepressants may have a beneficial role in its prevention. Newer antidepressants are less toxic and produce far fewer fatalities following overdoses than the older tricyclic and monoamine oxidase inhibitor antidepressants. As such, they are safer for people attempting suicide by acute overdose. Nevertheless, antidepressants as a class have not been associated with lower suicide risk among depressed persons, possibly because of their potential to worsen agitation or activation and induce mixed manic–depressive

https://doi.org/10.3390/ph18020258

features [5,6]. Depressive episodes in bipolar disorders, especially with dysphoric–mixed features, are far more likely to be associated with suicidal risk than hypomania or mania ("[hypo]mania") [7,8]. This finding is consistent with a reported protective action against suicidal behavior associated with hyperthymic affective temperament [9]. Long-term treatment with lithium has been associated with the prevention or amelioration of manic, depressive, and mixed episodes of bipolar disorder, and these clinical benefits may contribute to suicide prevention.

In 1997, we reviewed the effect of lithium treatment on suicidal behavior in major mood disorder based on findings from 28 studies (1974–1996) featuring 17,000 patients with mood disorders [10]. The rate of suicidal acts (attempts and suicides) during lithium treatment was 0.37%/year, compared to 3.20%/year without lithium treatment—an 8.65-fold difference. In 12 of these 28 reports [11–22] which compared the suicidal rates with and without lithium, the annual rates of suicidal acts were substantially lower during maintenance treatment with lithium. Despite the methodological heterogeneity of these reports, the rate of suicidal acts (±SD) among patients undergoing lithium treatment averaged 0.37 ± 0.66%/year (or per 100 patient-years) in 23 studies, compared to 3.39 ± 6.54%/year for patients without lithium treatment in 13 of the reports, corresponding to a 9.16-fold difference (paired-$t = 2.22$; $p = 0.03$).

Our review was preceded by a similar one [23], published in the now defunct journal *Lithium*, with similar findings. In the first studies hypothesizing a suicide prevention effect of lithium [24,25], in which the investigators interviewed relatives, witnesses, and medical records associated with 100 recent suicides using a questionnaire which included psychiatric and medical history. Of interest, 93/100 cases were considered mentally ill (70% with depression and 15% with alcoholism), and only 1 patient was treated with lithium. The authors proposed that, if all patients had been treated with lithium, an expected decrease in relapses would have avoided 21% of the suicides.

Among other studies in our 1997 review [10], we included a double-blind treatment trial [26] of 249 patients with bipolar disorder and 78 patients with unipolar depressive disorder, among whom 2 suicides were associated with the placebo treatment (1.28%/year) and none with the lithium treatment during 24 months of follow-up. Kay and Petterson [12] also found no suicides among 123 patients followed briefly in a specialized lithium clinic. Poole et al. [27], in a comparison of five years before versus during lithium treatment, found half as many suicide attempts during the treatment of 100 patients with various mood disorders. Lepkifker et al. [14] compared the rate of suicide in 33 patients with unipolar depressive disorder before versus during lithium maintenance treatment for an average of 8.30 years and found that 21.2% had attempted suicide before treatment (2.55%/year) compared to none during lithium treatment.

Coppen and colleagues [28] also found no suicides in 103 patients with major affective disorder attending a lithium clinic for 11 years (<0.088%/year). A later extension of this study identified 1 suicide in a total of 1519 patient-years (0.066%/year) of lithium maintenance treatment for major affective disorder [29]. Moreover, that study found a 13.8-fold higher suicide rate (0.91%/year) was found in 27 patients with untreated unipolar depressive disorder, suggesting a possible protective action of lithium in major depressive disorder [29].

Müller-Oerlinghausen and collaborators [15] designed a case–control study of suicidal risk in 68 patients with various major affective disorders and at least one suicide attempt, before, during an average of 8.00 years of lithium treatment, and following its discontinuation. They found 2.1 ± 0.2 suicide attempts before lithium treatment, 2 suicides and 4 attempts during treatment (suicidal acts: 1.10%/year), and 11 after lithium discontinuation (2.02%/year). Later, Nilsson [20] found a 4.80-times (95%CI: 1.10–12.6, $p = 0.02$) higher rate of suicides in patients who discontinued lithium compared to when they had been receiving lithium treatment.

In our study of 360 patients with type I or II bipolar disorder before, during, and following the discontinuation of long-term lithium monotherapy, the rates of suicide and life-threatening attempts were 6.40-fold lower during lithium treatment than either before lithium administration or long after lithium was discontinued [22]. Notably, however, the risk of suicidal acts increased by 20-fold within several months after discontinuing lithium maintenance treatment and later fell back to the same level encountered before lithium treatment had started. Moreover, the early suicidal risk following the discontinuation of long-term treatment with lithium was two times higher following abrupt or rapid as opposed to gradual discontinuation of lithium (over <14 vs. ≥14 days). These studies were included in a meta-analysis of 22 studies representing 5647 patients (33,473 patient-years of risk) among whom suicide was 81.8% less frequent during lithium treatment (0.159 vs. 0.875 deaths/100 patient-years), with a computed risk ratio in studies with rates off/on lithium of 8.85 (CI: 4.12–19.1, $p < 0.0001$) [3].

In a later meta-analysis of 31 clinical reports (including 5 randomized, controlled trials in which suicidal behavior was reported incidentally as an adverse event rather than as a specified outcome), we found a 4.91-fold lower risk of suicides and attempts in patients with recurrent major mood disorders treated with lithium [30].

2. Further Studies of Lithium Treatment and Suicidal Risk

In a meta-analysis of eight studies of patients with unipolar recurrent major depression, we found that long-term lithium treatment was again associated with a substantial reduction in the risk of suicides and attempts (by approximately 76%) among patients treated with lithium compared to alternatives, mainly anticonvulsants [31]. Three more recent reviews [32–34] added further support for a beneficial effect of lithium on suicidal behavior in major depressive disorder.

Studies supporting a beneficial effect of lithium in reducing suicidal risk include eight randomized controlled trials, in which suicidal behavior was again reported among the adverse events and not as an explicit, predefined outcome measure [35]. In those trials, the risk of suicide attempts and suicides was 1130/100 k PEY (100,000 person-exposure years; CI: 619–1889) among patients treated with the placebo or an alternative mood-stabilizer versus 119/100 k PEY (CI: 3.01–661) with lithium treatment, indicating a 9.50-fold difference ($\chi^2 = 7.15$, $p = 0.0075$). In addition, a rare randomized, placebo-controlled study with suicidal behavior as an explicit outcome measure [36] found a substantial but statistically nonsignificant difference in the rates of suicidal acts between patients with major affective disorder treated for 12 months with lithium versus a placebo, whereby all three observed suicides occurred in the placebo-treated group. Also, Koek et al. [37] found, in a study based on retrospective data from a sample of 161 military veterans diagnosed with bipolar disorder, that the rates of nonlethal suicide-related events were 9.67-times lower during long-term lithium treatment than with other mood-stabilizers given with antipsychotic drugs.

To date, reduced suicidal risk during long-term treatment with lithium in patients with bipolar disorder or recurrent major affective disorder has been found in 35 original studies included in reviews and meta-analyses [3,30,38–45]. In addition, a large study based on more than 30,000 subjects (age > 15 years) confirmed a lower rate of suicide and self-harm associated with lithium treatment [46], similarly to a Taiwanese nationwide cohort survey [47], based on nearly 26,000 participants with bipolar disorder, in which suicide accounted for 19.0% of mortality with an SMR of 26.0-times above the suicide rates in the general population. Lithium was highly significantly associated with a more than two-fold lower risk of suicide than anticonvulsants (carbamazepine, lamotrigine, or valproate) [47].

In a recent study [45], we analyzed data from a total of 13 randomized, controlled trials (RCTs) reported from 1973 to 2022. Participants had a diagnosis of either bipolar disorder or a variety of recurrent major affective disorders, excluding studies involving only unipolar major depression and those considered uninformative as lacking data on suicidal events with either treatment type. Again, suicidal behavior was not an explicit outcome measure but rather was noted among adverse events. The study involved 3836 subjects during treatment with lithium alone compared to the placebo (n = 1498) or a mood-stabilizing anticonvulsant or modern antipsychotic (n = 2338) [11,31,46–56].

The crude pooled rate of suicides and attempts among patients treated with lithium was 14/1498, or 0.935% [CI: 0.512–1.56], as opposed to 36/2338, or 1.54% [1.08–2.13], with alternative treatments. After correcting for a mean exposure time of 1.73 years in both groups, the rate per 100,000 person-exposure years (100k PEY) averaged 540 among patients treated with lithium versus 890 for those treated with the placebo or other treatments, indicating a 1.39-fold lower risk of suicidal behaviors after lithium administration. Data from the same 13 RCTs were also subjected to a meta-analysis (Figure 1), yielding an overall odds ratio of 0.491 [CI: 0.278–0.864], favoring lithium (z-score = 2.46, p = 0.01).

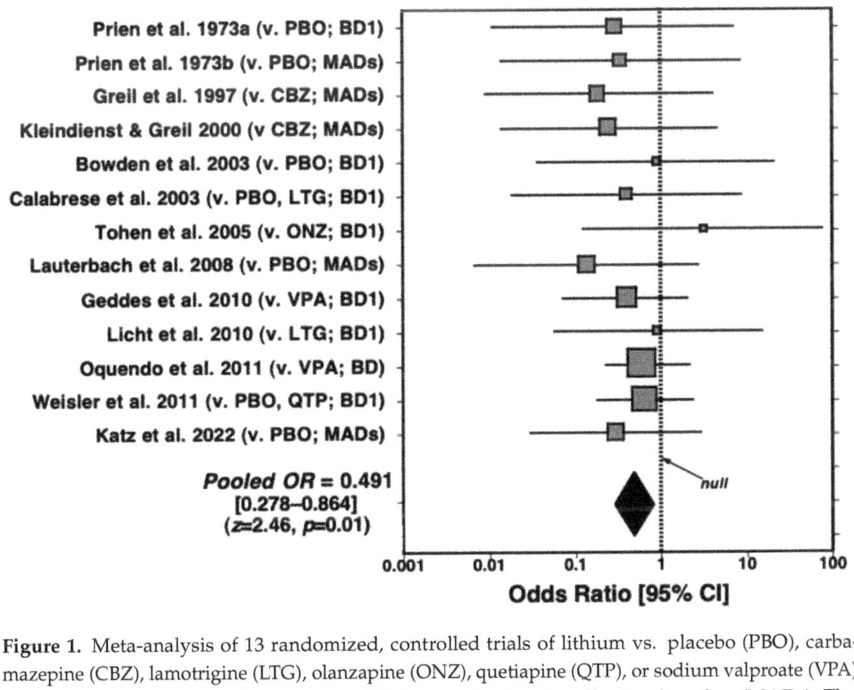

Figure 1. Meta-analysis of 13 randomized, controlled trials of lithium vs. placebo (PBO), carbamazepine (CBZ), lamotrigine (LTG), olanzapine (ONZ), quetiapine (QTP), or sodium valproate (VPA) given to patients with bipolar I disorder (BD1) or a mix of major affective disorders (MADs). The rates of suicide or attempts were significantly lower with lithium administration (pooled OR = 0.491; CI: 0.278–0.864; z-score = 2.46; and $p < 0.01$) [11,26,36,49–56]. From Tondo & Baldessarini [45].

3. Prevention of All-Cause Early Mortality with Lithium

Suicide is the leading cause of early death in people with bipolar disorder. However, these patients also show a tendency toward early mortality due to other illnesses, particularly cardiovascular and pulmonary diseases [57–60]. In several studies conducted since the early 1990s, lithium has been found effective in reducing early mortality from all causes [34,47,61–63] and is probably more effectively than treatment with antipsychotics or anticonvulsants [47,64]. Lithium's effectiveness in preventing early mortality has also been confirmed indirectly by an increase in the all-cause mortality after treatment discontinuation [65].

It is noteworthy that a reduction in early mortality due to suicide has been found despite the potentially lethal toxicity of lithium in cases of acute overdoses. Mortality from overdoses of several agents was recently ranked as follows: tricyclic antidepressants (40.7 deaths per 10,000 overdoses) > acetaminophen (25.8) > lithium (13.2) > modern antipsychotics (5.80) [66]. Lethal outcomes with acute overdoses of lithium can be avoided by vomiting or timely hemodialysis [67,68]. However, attempting suicide with lithium does not seem to be common [69], perhaps because it is not widely considered to be a lethal drug. It is also possible that suicide attempts using this agent are limited by its modulation of impulsive and aggressive behaviors [70].

4. Controversies About Suicide Prevention with Lithium Treatment

In the half-century of research on the potential anti-suicidal effect of lithium treatment, several ethical and methodological criticism have emerged. For instance, a randomized trial among 519 members of the US Veterans Administration (VA) was interrupted for futility after several months without suicide-related events, with rates remaining similar between the lithium and placebo treatments [56]. Rather than considering it as evidence of the failure of lithium treatment, the study should have been considered inconclusive owing to the following: [a] low serum lithium levels; [b] the use of additional uncontrolled treatments; [c] the brief duration of lithium treatment; [d] the exclusion of patients at relatively high suicidal risk; and [e] the heterogeneity of diagnoses, including unipolar depression and co-occurring substance abuse [35,71]. Of note, within one month of the end of the trial, three more suicides occurred in the placebo group and were not included in the trial data analysis. However, even with these suicides included, any effects of lithium remained nonsignificant. We commented that the trial by Katz et al. [56] did not find evidence of an anti-suicidal effect because lithium was added to other treatment regimens for groups comprising a relatively small number of mostly male veterans with complex psychopathological conditions, involving relatively brief treatment with low circulating concentrations of lithium, supporting the conclusion that their findings cannot be taken as evidence that lithium lacks anti-suicidal effects [35].

In another recent negative review of evidence on the anti-suicidal effect of lithium treatment [72], the authors found no difference in the risk of suicide between the treatment with lithium or the placebo in 12 RCTs with a total of 2578 participants. A reaction by the International Group for The Study of Lithium Treated Patients (IGSLI) [73] criticized the validity of the above review, underscoring [a] the arbitrary exclusion of trials carried out before the year 2000, [b] the selective inclusion of only some studies from an earlier meta-analysis [39], [c] the exclusion of studies on lithium versus active comparators, and [d] the inclusion of studies with no information on suicide, under the questionable assumption that there were none. Comments by Bschor et al. [73] stirred another reaction by the authors of the original review [74], claiming that the exclusion of studies before the year 2000 was justified on the basis of the unreliability of their data, whilst reporting that a sensitivity analysis also including pre-2000 data produced the same negative results. They also addressed their choice of not comparing lithium with an active comparator but only against a placebo, without providing a plausible rationale.

Evidence to test the hypothesis that lithium or other treatments may reduce suicidal risk is difficult to gather. Particularly rare are randomized controlled trials against either a placebo or an alternative active treatment with suicide-associated behavior as the defined outcome, similarly to the studies performed to support the superiority of clozapine over olanzapine in the treatment of patients with psychotic disorder and high suicidal risk, based on surrogate measures such as interventions to prevent suicide rather than suicide attempts or deaths [75,76]. Making potential fatalities an explicit outcome measure faces severe ethical and practical considerations, but surrogate outcomes such as suicidal ideation or interventions intended to prevent suicide may not be reliable substitutes for actual suicidal behavior.

Again, almost all research on treatment effects on suicidal behavior has been based on incidental reports of suicidal behavior as an adverse event rather than an explicit outcome measure. This method raises the possibility of underestimating suicidal events. In addition, such study designs are not considered adequate for the regulatory support of a therapeutic indication for a specific anti-suicidal effect.

Additional matters of concern arise in studies comparing the presence of lithium treatment to its absence, including the mismatching of other clinical interventions which may alter suicidal risk. In studies involving continued versus discontinued lithium treatment, the discontinuation itself may contribute to an elevated suicidal risk [22,77].

5. Lithium in Drinking Water

We found 42 research reports addressing the relationship of lithium concentrations in drinking water to local rates of suicide or violent behavior [78–119]. Of these reports, 27/42 (64.3%; indicated with an asterisk in Table 1) reported data and 30/42 (71.4%) were based on ecological studies (13 from Europe, 9 from Japan, 6 from the USA, and 1 each from Chile and Iran). Of all 42 studies, 27 of them reported data on the correlations between suicide rates or SMRs and the lithium concentration in local water. Of these, 13 (48.1%) found an inverse relationship, 8 (29.6%) a partial correlation, and 6 (22.2%) no correlation. Among the 10 reviews on this topic, including the studies already cited above, 5 (50%) reported a partial correlation, 4 (40%) an inverse correlation, and 1 (10%) no correlation (Table 1).

Table 1. Correlation between lithium in water and suicide or violence.

Study	Year	Country	Study Type	Results	Comment
Dawson et al. [78] *	1972	USA	Ecological	Inverse correlation	Reduced hospitalization and violence
Schrauzer & Shrestha [79] *	1990	USA	Ecological	Inverse correlation	Suicides and crimes; favors Li supplementation
Schrauzer [80]	2002		Commentary	No correlation	Lithium in food
Ohgami et al. [81] *	2009	Japan	Ecological	Inverse correlation	——
Kabacs et al. [82] *	2011	England	Ecological	No correlation	——
Kapusta et al. [83] *	2011	Austria	Ecological	Inverse correlation	——
Blüml et al. [84] *	2013	USA	Ecological	Inverse correlation	——

Table 1. *Cont.*

Study	Year	Country	Study Type	Results	Comment
Giotakos et al. [85] *	2013	Greece	Ecological	Partial correlation	Lower rates of homicides, rapes, and substance abuse
Sugawara et al. [86] *	2013	Japan	Ecological	Partial correlation	Only in women
Huber et al. [87]	2014	USA	Review	——	Varied with altitude
Giotakos et al. [88] *	2015	Greece	Ecological	Partial correlation	Homicides
Ishii et al. [89] *	2015	Japan	Ecological	Partial correlation	Only in males
Pompili et al. [90] *	2015	Italy	Ecological	No correlation	——
Vita et al. [91]	2015	——	Review	Partial correlation	——
Goldstein & Mascitelli [92]	2016	——	Review	Inverse correlation	Add Li to prevent violence
Shiotsuki et al. [93] *	2016	Japan	Ecological	Partial correlation	In men only and associated with the climate
Ando et al. [94] *	2017	Japan	Cohort	Inverse correlation	In adolescents, less depression and aggressiveness
Awad et al. [95]	2017	——	Commentary	——	Against adding lithium to water
Kanehisa et al. [96] *	2017	Japan	Case–control	Inverse correlation	Fewer suicide attempts
Knudsen et al. [97] *	2017	Denmark	Cohort study	No correlation	——
König et al. [98] *	2017	Chile	Ecological	Inverse correlation	Only in some regions
Brown et al. [99]	2018	——	Review	Partial correlation	May protect from lead exposure
Fajardo et al. [100] *	2018	USA	Ecological	Inverse correlation	Decreased all-cause mortality and loss of years
Kurosawa et al. [101] *	2018	Japan	Case–control	Inverse correlation	Lithium added to omega-3 fatty acids
Oliveira et al. [102] *	2019	Portugal	Ecological	No correlation	——
Palmer et al. [103] *	2019	USA	Ecological	Inverse correlation	——

Table 1. *Cont.*

Study	Year	Country	Study Type	Results	Comment
Barjasteh-Askari et al. [104]	2020	——	Review	Partial correlation	Dose-dependent effect
Del Matto et al. [105]	2020	——	Review	Partial correlation	More in long-term lithium treatment
Eyre-Watt et al. [106]	2020	——	Review	Inverse correlation	——
Kozaka et al. [107] *	2020	Japan	Ecological	No correlation	——
Memon et al. [108]	2020	——	Review	Inverse correlation	——
Kugimiya et al. [109] *	2021	Japan	Ecological	Partial correlation	Only in men
Liaugaudaite et al. [110] *	2021	Lithuania	Ecological	Inverse correlation	——
Araya et al. [111]	2022	——	Commentary	——	Proposed a consensus
Izsak et al. [112] *	2022	Hungary	Ecological	Partial correlation	Only in women
Kawada [113]	2022	——	Commentary	——	——
Liaugaudaite et al. [114] *	2022	Lithuania	Ecological	Partial correlation	Only with higher lithium levels and high rates of affective disorders
Fadaei [115]	2023	——	Review	Inverse correlation	——
Rihmer & Dome [115]	2023	Hungary	——	——	In favor of the fortification of drinking water
Bjørklund & Storchylo [116]	2024	——	Review	Partial correlation	Critical comments on lithium supplementation
Harandi et al. [117] *	2024	Iran	Ecological	Inverse correlation	Reduced suicide attempts
Pichler et al. [118] *	2024	Switzerland	Ecological	No correlation	——

Of the 27/42 studies reporting data (*), 13 (48.1%) found an inverse relationship between suicide rates or SMRs and lithium concentration in local water, 8 (29.6%) found partial correlations, and 6 (22.2%) found no correlation. Among the 10 reviews, including other listed studies, 5 (50%) reported a partial correlation, 4 (40%) an inverse correlation, and 1 (10%) no correlation.

The significance of these findings is unclear, particularly considering the extremely low concentrations of lithium in drinking water compared to the doses and blood concentrations commonly used for treating mood disorders [67]. However, it might be that exposure to even small doses of lithium during crucial times in postnatal development or for prolonged periods may have significant effects. In addition, ingested water is not just what is drunk but also what is contained in foods [79]. Rigorous studies on this topic should include only persons always residing in the given area and consistently drinking tap water rather than bottled water. Despite inconsistencies in the studies, some investigators have been sufficiently convinced to propose the addition of lithium to drinking water, much like fluoride was added to water to reduce the risk of dental caries [107,110,115].

6. Conclusions

The research evidence reviewed here provides strong support for an anti-suicidal effect of long-term treatment with lithium, at least in patients with bipolar disorder and probably also in those with major depressive disorder. Moreover, studies concerning the anti-suicidal effects of lithium include evidence of its superiority to alternative treatments, including mood-altering anticonvulsants (notably, carbamazepine, lamotrigine, and sodium valproate) and both first- and second-generation antipsychotics. The relevant evidence reviewed above includes many clinical comparisons of the rates of suicides and attempts during treatment with lithium or other agents, as well as evidence from at least eight randomized, blinded, controlled trials comparing lithium with a placebo or other mood-stabilizing medicines in which suicidal behavior was noted incidentally as an adverse outcome rather than as an explicit, planned outcome measure. Studies designed with suicidal behavior as a stated outcome are rare, as they are both ethically and practically challenging, and are particularly unlikely with lithium, which has received very limited commercial support as an unpatentable mineral.

The mechanisms of lithium's anti-suicidal effects could be associated with a direct action against aggressiveness and impulsivity, which are invariably present in suicide behaviors, or the indirect benefit of lithium in preventing mood states, especially depressive states with mixed features [8]. It has been proposed that the anti-aggressive effect can be obtained with small doses of lithium, as with highly lithiated drinking water [120], whereas the mood-stabilizing effects require therapeutic doses. The lengthy process highlighting the anti-suicidal effect of lithium in patients with bipolar disorder has certainly not been without limitations. The most significant, as previously noted, concerns the incidental nature of suicidal events in various studies in which they were considered adverse effects of the therapies. Other limitations include the different designs of the studies (randomized controlled trials or open studies), the duration of the treatments, and the lack of uniformity in the therapies compared to lithium salts. Additionally, the nomenclature associated with suicide introduces variability, as suicidal acts themselves can differ between, for instance, violent or non-violent acts, attempts with clear intent or without, and methods which are more or less lethal. Nonetheless, it is noteworthy that, despite all these potential confounding sources, even the most recent studies have consistently demonstrated the efficacy of long-term treatment with lithium salts in patients with bipolar disorder, as also shown in patients with schizophrenia. Ideally, future research on suicide prevention in patients with bipolar disorder should be conducted using a randomized controlled design, including a large number of patients followed over a sufficiently long period of time. However, these requirements face significant challenges, as suicidal behaviors are rare events and the above-proposed studies would need to span several years. For this reason, securing funding would be nearly impossible, in addition to the lack of marketing interest in lithium therapy.

Author Contributions: Conceptualization, L.T. & R.J.B.; methodology, L.T. & R.J.B.; investigation, L.T. & R.J.B.; writing—original draft preparation, L.T.; writing—review and editing, L.T. & R.J.B. All authors have read and agreed to the published version of the manuscript.

Funding: This research received no external funding.

Informed Consent Statement: Not applicable.

Data Availability Statement: Not applicable.

Conflicts of Interest: The authors declare no conflict of interest.

References

1. Goodwin, F.K.; Jamison, K.R. *Manic-Depressive Illness*; Oxford University Press: New York, NY, USA, 1990; pp. 227–244, 630–724.
2. Harris, E.C.; Barraclough, B. Suicide as an outcome for mental disorders: A meta-analysis. *Br. J. Psychiatry* **1997**, *170*, 205–228. [CrossRef]
3. Tondo, L.; Hennen, J., Baldessarini, R.J. Reduced suicide risk with long-term lithium treatment in major affective Illness: A meta-analysis. *Acta Psychiatr. Scand.* **2001**, *104*, 163–172. [CrossRef]
4. Jacobs, D.G.; Klein-Benheim, M. Current Status of Suicide-Focused Assessment and Treatment. 2024. Available online: https://www.mcleanhospital.org/suicide-assessment (accessed on 30 November 2024).
5. Culpepper, L.; Davidson, J.R.; Dietrich, A.J.; Goodman, W.K.; Kroenke, K.; Schwenk, T.L. Suicidality as a possible side effect of antidepressant treatment. *J. Clin. Psychiatry* **2004**, *65*, 742–749. [CrossRef] [PubMed]
6. Nierenberg, A.A.; Agustini, B.; Köhler-Forsberg, O.; Cusin, C.; Katz, D.; Sylvia, L.G.; Peters, A.; Berk, M. Diagnosis and Treatment of Bipolar Disorder: A Review. *JAMA* **2023**, *330*, 1370–1380. [CrossRef] [PubMed]
7. Swann, A.C.; Lafer, B.; Perugi, G.; Frye, M.A.; Bauer, M.; Bahk, W.M.; Scott, J.; Ha, K.; Suppes, T. Bipolar mixed states: An International Society for Bipolar Disorders task force report of symptom structure, course of illness, and diagnosis. *Am. J. Psychiatry* **2013**, *170*, 31–42. [CrossRef]
8. Tondo, L.; Vázquez, G.H.; Baldessarini, R.J. Suicidal behavior associated with mixed features in major mood disorders. *Psychiatr. Clin. N. Am.* **2020**, *43*, 83–93. [CrossRef] [PubMed]
9. Tondo, L.; Vázquez, G.H.; Sani, G.; Pinna, M.; Baldessarini, R.J. Association of suicidal risk with ratings of affective temperaments. *J. Affect. Disord.* **2018**, *229*, 322–327. [CrossRef]
10. Tondo, L.; Jamison, K.R.; Baldessarini, R.J. Effect of lithium maintenance on suicidal behavior in major mood disorders. *Ann. N. Y Acad. Sci.* **1997**, *836*, 339–351. [CrossRef]
11. Prien, R.F.; Klett, C.J.; Caffey, E.M., Jr. Lithium carbonate and imipramine in prevention of affective episodes: A comparison in recurrent affective illness. *Arch. Gen. Psychiatry* **1973**, *29*, 420–425. [CrossRef]
12. Kay, D.W.K.; Petterson, U. Manic-depressive illness. *Acta Psychiatr. Scand.* **1977**, *269*, 55–60. [CrossRef]
13. Venkoba-Rao, A.; Hariharasubramanian, N.; Parvathi-Devi, S.; Sugumar, A.; Srinivasan, V. Lithium prophylaxis in affective disorder. *Indian. J. Psychiatry* **1982**, *23*, 22–30.
14. Lepkifker, E.; Horesh, N.; Floru, S. Long-term lithium prophylaxis in recurrent unipolar depression. A controversial indication? *Acta Psychiatr. Belg.* **1985**, *85*, 434–443. [PubMed]
15. Müller-Oerlinghausen, B.; Müser-Causemann, B.; Volk, J. Suicides and parasuicides in a high-risk patient group on and off lithium long-term medication. *J. Affect. Disord.* **1992**, *25*, 261–269. [CrossRef]
16. Rihmer, Z.; Szántó, K.; Barsi, J. Suicide prevention: Fact or fiction. *Br. J. Psychiatry* **1993**, *162*, 130–131. [CrossRef] [PubMed]
17. Coppen, A. Depression as a lethal disease: Prevention strategies. *J. Clin. Psychiatry* **1994**, *55*, 37–45. [PubMed]
18. Felber, W.; Kyber, A. Suizide und Parasuizide während und ausserhalb einer Lithium prophylaxe. In *Ziele und Ergebnisse der Medikamentosen Prophylaxe Affektiver Psychosen*; Müller-Oerlinghausen, B., Berghofer, A., Eds.; Thieme Verlag: Stuttgart, Germany, 1994; pp. 53–59.
19. Lenz, G. Increased mortality after drop-out from lithium clinic. *Neuropsychopharmacology* **1994**, *10*, 682s.
20. Nilsson, A. Mortality in recurrent mood disorders during periods on and off lithium: A complete population study in 362 patients. *Pharmacopsychiatry* **1995**, *28*, 8–13. [CrossRef]
21. Thies-Flechtner, K.; Müller-Oerlinghausen, B.; Seibert, W.; Walther, A.; Greil, W. Effect of prophylactic treatment on suicide risk in patients with major affective disorders (Data from a randomized prospective trial). *Pharmacopsychiatry* **1996**, *29*, 103–107. [CrossRef] [PubMed]
22. Tondo, L.; Baldessarini, R.J.; Hennen, J.; Floris, G.; Silvetti, F.; Tohen, M. Lithium treatment and risk of suicidal behavior in bipolar disorder patients. *J. Clin. Psychiatry* **1998**, *59*, 405–414. [CrossRef]
23. Crundwell, J.K. Lithium and its potential benefit in reducing increased mortality rates due to suicide. *Lithium* **1994**, *5*, 193–204.
24. Barraclough, B. Suicide prevention, recurrent affective disorder and lithium. *Br. J. Psychiatry* **1972**, *121*, 391–392. [CrossRef] [PubMed]
25. Barraclough, B.; Bunch, J.; Nelson, B.; Sainsbury, P. A hundred cases of suicide: Clinical aspects. *Br. J. Psychiatry* **1974**, *125*, 355–373. [CrossRef] [PubMed]
26. Prien, R.F.; Point, P.; Caffey, E.M., Jr.; Klett, C.J. Prophylactic efficacy of lithium carbonate in manic-depressive illness: Report of the Veterans Administration and National Institute of Mental Health collaborative study group. *Arch. Gen. Psychiatry* **1974**, *28*, 337–341. [CrossRef] [PubMed]

27. Poole, A.J.; James, H.D.; Hughes, W.C. Treatment experiences in the lithium clinic at St Thomas Hospital. *J. R. Soc. Med.* **1978**, *71*, 890–894. [CrossRef]
28. Coppen, A.; Standish-Barry, H.; Bailey, J.; Houston, G.; Silcocks, P.; Hermon, C. Does lithium reduce the mortality of recurrent mood disorders? *J. Affect. Disord.* **1991**, *23*, 1–7. [CrossRef] [PubMed]
29. Coppen, A.; Farmer, R. Suicide mortality in patients on lithium maintenance therapy. *J. Affect. Disord.* **1998**, *50*, 261–267. [CrossRef] [PubMed]
30. Baldessarini, R.J.; Tondo, L.; Davis, P.; Pompili, M.; Goodwin, F.K.; Hennen, J. Decreased suicidal risk during long-term lithium treatment: Meta-analysis. *Bipolar Disord.* **2006**, *8*, 625–639. [CrossRef] [PubMed]
31. Guzzetta, F.; Tondo, L.; Centorrino, F.; Baldessarini, R.J. Lithium treatment reduces suicide risk in recurrent major depressive disorder. *J. Clin. Psychiatry* **2007**, *68*, 380–383. [CrossRef] [PubMed]
32. Cipriani, A.; Hawton, K.; Stockton, S.; Geddes, J.R. Lithium in the prevention of suicide in mood disorders: Updated systematic review and meta-analysis. *BMJ* **2013**, *346*, f3646–f3658. [CrossRef]
33. Turecki, G.; Brent, D.A. Suicide and suicidal behaviour. *Lancet* **2016**, *387*, 1227–1239. [CrossRef] [PubMed]
34. Abou-Saleh, M.; Müller-Oerlinghausen, B.; Coppen, A.J. Lithium in the episode and suicide prophylaxis and in augmenting strategies in patients with unipolar depression. *Int. J. Bipolar Disord.* **2017**, *5*, 11–19. [CrossRef] [PubMed]
35. Baldessarini, R.J.; Tondo, L. Testing for antisuicidal effects of lithium treatment. *JAMA Psychiatry* **2022**, *79*, 9–10. [CrossRef]
36. Lauterbach, E.; Felber, W.; Müller-Oerlinghausen, B.; Bronisch, T.; Meyer, T.; Kilb, B.; Lewitzka, U.; Hawellek, B.; Quante, A.; Richter, K.; et al. Adjunctive lithium treatment in the prevention of suicidal behaviour in depressive disorders: A randomised, placebo-controlled, 1-year trial. *Acta Psychiatr. Scand.* **2008**, *118*, 469–479. [CrossRef] [PubMed]
37. Koek, R.J.; Yerevanian, B.L.; Mintz, J. Subtypes of antipsychotics and suicidal behavior in bipolar disorder. *J. Affect. Disord.* **2012**, *143*, 27–33. [CrossRef]
38. Angst, J.; Angst, F.; Gerber-Werder, R.; Gamma, A. Suicide in 406 mood-disorder patients with and without long-term medication: 40–44 years follow-up. *Arch. Suicide Res.* **2005**, *9*, 279–300. [CrossRef]
39. Cipriani, A.; Pretty, H.; Hawton, K.; Geddes, J.R. Lithium in the prevention of suicidal behavior and all-cause mortality in patients with mood disorders: Systematic review of randomized trials. *Am. J. Psychiatry* **2005**, *162*, 1805–1819. [CrossRef] [PubMed]
40. Kessing, L.V.; Søndergard, L.; Kvist, K.; Andersen, P.K. Suicide risk in patients treated with lithium. *Arch. Gen. Psychiatry* **2005**, *62*, 860–866. [CrossRef] [PubMed]
41. Müller-Oerlinghausen, B.; Ahrens, B.; Felber, W. The suicide-preventive and mortality-reducing effect of lithium. In *Lithium in Neuropsychiatry*; Bauer, M., Grof, P., Müller-Oerlinghausen, B., Eds.; Informa Healthcare: London, UK, 2013; pp. 179–192.
42. Riblet, N.B.; Shiner, B.; Young-Xu, Y.; Watts, B.V. Lithium in the prevention of suicide in adults: Systematic review and meta-analysis of clinical trials. *Br. J. Psychiatry Open* **2022**, *8*, e199–e202. [CrossRef] [PubMed]
43. Wilkinson, S.T.; Trujillo Diaz, D.; Rupp, Z.W.; Kidambi, A.; Ramirez, K.L.; Flores, J.M.; Avila-Quintero, V.J.; Rhee, T.G.; Olfson, M.; Bloch, M.H. Pharmacological and somatic treatment effects on suicide in adults: A systematic review and meta-analysis. *Depress. Anxiety* **2022**, *39*, 100–112. [CrossRef]
44. Pompili, M.; Berardelli, I.; Sarubbi, S.; Rogante, E.; Germano, L.; Sarli, G.; Erbuto, D.; Baldessarini, R.J. Lithium treatment versus hospitalization in bipolar disorder and major depression patients. *J. Affect. Disord.* **2023**, *340*, 245–249. [CrossRef] [PubMed]
45. Tondo, L.; Baldessarini, R.J. Prevention of suicidal behavior with lithium treatment in patients with recurrent mood disorders. *Int. J. Bipolar Disord.* **2024**, *12*, 6–19. [CrossRef] [PubMed]
46. Fitzgerald, C.; Christensen, R.H.B.; Simons, J.; Andersen, P.K.; Benros, M.E.; Nordentoft, M.; Erlangsen, A.; Hawton, K. Effectiveness of medical treatment for bipolar disorder regarding suicide, self-harm and psychiatric hospital admission: Between- and within-individual study on Danish national data. *Br. J. Psychiatry* **2022**, *221*, 692–700. [CrossRef] [PubMed]
47. Chen, P.H.; Tsai, S.Y.; Chen, P.Y.; Pan, C.H.; Su, S.S.; Chen, C.C.; Kuo, C.J. Mood stabilizers and risk of all-cause, natural, and suicide mortality in bipolar disorder: A nationwide cohort study. *Acta Psychiatr. Scand.* **2023**, *147*, 234–247. [CrossRef] [PubMed]
48. Greil, W.; Ludwig-Mayerhofer, W.; Erazo, N.; Schöchlin, C.; Schmidt, S.; Engel, R.R.; Czernik, A.; Giedke, H.; Müller-Oerlinghausen, B.; Osterheider, M.; et al. Lithium versus carbamazepine in the maintenance treatment of bipolar disorders—A randomised study. *J. Affect. Disord.* **1997**, *43*, 151–161. [CrossRef] [PubMed]
49. Kleindienst, N.; Greil, W. Differential efficacy of lithium and carbamazepine in the prophylaxis of bipolar disorder: Results of the MAP study. *Neuropsychobiology* **2000**, *42* (Suppl. S1), 2–10. [CrossRef]
50. Bowden, C.L.; Calabrese, J.R.; Sachs, G.; Yatham, L.N.; Asghar, S.A.; Hompland, M.; Montgomery, P.; Earl, N.; Smoot, T.M.; DeVeaugh-Geiss, J.; et al. A placebo-controlled 18-month trial of lamotrigine and lithium maintenance treatment in recently manic or hypomanic patients with bipolar I disorder. *Arch. Gen. Psychiatry* **2003**, *60*, 392–400. [CrossRef]
51. Geddes, M.; Greil, W.; Calabrese, J.R.; Sachs, G.S.; Yatham, L.N.; Oerlinghausen, B.M.; Koukopoulos, A.; Cassano, G.B.; Grunze, H.; Licht, R.W.; et al. Olanzapine versus lithium in the maintenance treatment of bipolar disorder: A 12-month, randomized, double-blind, controlled clinical trial. *Am. J. Psychiatry* **2005**, *162*, 1281–1290.

52. Geddes, J.R.; Goodwin, G.M.; Rendell, J.; Azorin, J.M.; Cipriani, A.; Ostacher, M.J.; Alder, N.; Morriss, R.; Juszczak, E. (BALANCE investigators and collaborators). Lithium plus valproate combination therapy versus monotherapy for relapse prevention in bipolar I disorder: A randomised open-label trial. *Lancet* **2010**, *375*, 385–395.
53. Licht, R.W.; Nielsen, J.N.; Gram, L.F.; Vestergaard, P.; Bendz, H. Lamotrigine versus lithium as maintenance treatment in bipolar I disorder: An open, randomized effectiveness study mimicking clinical practice. The 6th trial of the Danish University Antidepressant Group (DUAG-6). *Bipolar Disord.* **2010**, *12*, 483–493. [CrossRef]
54. Oquendo, M.A.; Galfalvy, H.C.; Currier, D.; Grunebaum, M.F.; Sher, L.; Sullivan, G.M.; Burke, A.K.; Harkavy-Friedman, J.; Sublette, M.E.; Parsey, R.V.; et al. Treatment of suicide attempters with bipolar disorder: A randomized clinical trial comparing lithium and valproate in the prevention of suicidal behavior. *Am. J. Psychiatry* **2011**, *168*, 1050–1056. [CrossRef] [PubMed]
55. Weisler, R.H.; Nolen, W.A.; Neijber, A.; Hellqvist, Š.; Paulsson, B. Continuation of quetiapine versus switching to placebo or lithium for maintenance treatment of bipolar I disorder (Trial 144: A randomized controlled study). *J. Clin. Psychiatry* **2011**, *72*, 1452–1464. [CrossRef] [PubMed]
56. Katz, I.R.; Rogers, M.P.; Lew, R.; Thwin, S.S.; Doros, G.; Ahearn, E.; Ostacher, M.J.; DeLisi, L.E.; Smith, E.G.; Ringer, R.J.; et al. Lithium Treatment in the Prevention of Repeat Suicide-Related Outcomes in Veterans With Major Depression or Bipolar Disorder: A Randomized Clinical Trial. *JAMA Psychiatry* **2022**, *79*, 24–32. [CrossRef] [PubMed]
57. Ahrens, B.; Müller-Oerlinghausen, B.; Schou, M.; Wolf, T.; Alda, M.; Grof, E.; Grof, P.; Lenz, G.; Simhandl, C.; Thau, K.; et al. Excess cardiovascular and suicide mortality of affective disorders may be reduced by lithium prophylaxis. *J. Affect. Disord.* **1995**, *33*, 67–75. [CrossRef] [PubMed]
58. Sharma, R.; Markar, H.R. Mortality in affective disorder. *J. Affect. Disord.* **1994**, *31*, 91–96. [CrossRef] [PubMed]
59. Roshanaei-Moghaddam, B.; Katon, W. Premature mortality from general medical illnesses among persons with bipolar disorder: A review. *Psychiatr. Serv.* **2009**, *60*, 147–156. [CrossRef] [PubMed]
60. De Hert, M.; Correll, C.U.; Bobes, J.; Cetkovich-Bakmas, M.; Cohen, D.; Asai, I.; Detraux, J.; Gautam, S.; Möller, H.J.; Ndetei, D.M.; et al. Physical illness in patients with severe mental disorders. I. Prevalence, impact of medications and disparities in health care. *World Psychiatry* **2011**, *10*, 52–77. [CrossRef] [PubMed]
61. Müller-Oerlinghausen, B.; Ahrens, B.; Volk, J.; Grof, P.; Grof, E.; Schou, M.; Vestergåard, P.; Lenz, G.; Simhandl, C.; Thau, K. Reduced mortality of manic-depressive patients in long-term lithium treatment: An international collaborative study by IGSLI. *Psychiatry Res.* **1991**, *36*, 329–331. [CrossRef]
62. Toffol, E.; Hätönen, T.; Tanskanen, A.; Lönnqvist, J.; Wahlbeck, K.; Joffe, G.; Tiihonen, J.; Haukka, J.; Partonen, T. Lithium is associated with decrease in all- cause and suicide mortality in high-risk bipolar patients: A nationwide registry-based prospective cohort study. *J. Affect. Disord.* **2015**, *183*, 159–165. [CrossRef]
63. Carvalho, A.F.; Hsu, C.W.; Vieta, E.; Solmi, M.; Marx, W.; Berk, M.; Liang, C.S.; Tseng, P.T.; Wang, L.J. Mortality and lithium-protective effects after first-episode mania diagnosis in bipolar disorder: Nationwide retrospective cohort study in Taiwan. *Psychother. Psychosom.* **2024**, *93*, 36–45. [CrossRef] [PubMed]
64. Chan, J.K.N.; Wong, C.S.M.; Fang, C.Z.; Hung, S.C.; Lo, H.K.Y.; Chang, W.C. Mortality risk and mood stabilizers in bipolar disorder: A propensity-score-weighted population-based cohort study in 2002–2018. *Epidemiol. Psychiatr. Sci.* **2024**, *33*, e31–e40. [CrossRef]
65. Müller-Oerlinghausen, B.; Wolf, T.; Ahrens, B.; Glaenz, T.; Schou, M.; Grof, E.; Grof, P.; Lenz, G.; Simhandl, C.; Thau, K. Mortality of patients who dropped out from regular lithium prophylaxis: A collaborative study by the International Group for the Study of Lithium-treated patients (IGSLI). *Acta Psychiatr. Scand.* **1996**, *94*, 344–347. [CrossRef] [PubMed]
66. Nelson, J.C.; Spyker, D.A. Morbidity and mortality associated with medications used in the treatment of depression: Analysis of cases reported to U.S. Poison Control Centers, 2000–2014. *Am. J. Psychiatry* **2017**, *174*, 438–450. [CrossRef]
67. Watson, W.A.; Litovitz, T.L.; Klein-Schwartz, W.; Rodgers, G.C., Jr.; Youniss, J.; Reid, N.; Rouse, W.G.; Rembert, R.S.; Borys, D. 2003 annual report of the American Association of Poison Control Centers toxic exposure surveillance system. *Am. J. Emerg. Med.* **2004**, *22*, 335–404. [CrossRef] [PubMed]
68. Baldessarini, R.J. *Chemotherapy in Psychiatry*, 3rd ed.; Springer Press: New York, NY, USA, 2013.
69. Waddington, D.; McKensie, I.P. Overdose rates in lithium-treated vs. antidepressant-treated outpatients. *Acta Psychiatr. Scand.* **1994**, *90*, 50–52. [CrossRef]
70. Müller-Oerlinghausen, B.; Lewitzka, U. Lithium reduces pathological aggression and suicidality: A mini-review. *Neuropsychobiology* **2010**, *62*, 43–49. [CrossRef]
71. Manchia, M.; Sani, G.; Alda, M. Suicide risk and lithium. *JAMA Psychiatry* **2022**, *79*, 513–514. [CrossRef]
72. Nabi, Z.; Stansfeld, J.; Plöderl, M.; Wood, L.; Moncrieff, J. Effects of lithium on suicide and suicidal behaviour: A systematic review and meta-analysis of randomized trials. *Epidemiol. Psychiatr. Sci.* **2022**, *31*, e65–e75. [CrossRef] [PubMed]
73. Bschor, T.; Adli, M.; Alda, M.; Baethge, C.; Etain, B.; Glenn, T.; Grof, P.; Hajek, T.; Hayes, J.; Manchia, M.; et al. International Group for The Study of Lithium Treated Patients (IGSLI). Epidemiol. *Psychiatr. Sci.* **2022**, *31*, e84–e86.

74. Moncrieff, J.; Plöderl, M.; Nabi, Z.; Stansfeld, J.; Wood, L. Evidence does not confirm that lithium prevents suicide: A reply to Bschor et al. *Epidemiol. Psychiatr. Sci.* **2022**, *31*, e88–e89. [CrossRef] [PubMed]
75. Meltzer, H.Y.; Alphs, L.; Green, A.I.; Altamura, A.C.; Anand, R.; Bertoldi, A.; Bourgeois, M.; Chouinard, G.; Islam, M.Z.; Kane, J.; et al. Clozapine treatment for suicidality in schizophrenia: International Suicide Prevention Trial (InterSePT). *Arch. Gen. Psychiatry* **2003**, *60*, 82–91. [CrossRef]
76. Hennen, J.; Baldessarini, R.J. Suicidal risk during treatment with clozapine: A meta-analysis. *Schizophr. Res.* **2005**, *73*, 139–145. [CrossRef] [PubMed]
77. Tondo, L.; Baldessarini, R.J. Discontinuing psychotropic drug treatment. *BJPsych Open* **2020**, *6*, e24. [CrossRef] [PubMed]
78. Dawson, E.B.; Moore, T.D.; McGanity, W.J. Relationship of lithium metabolism to mental hospital admissions and homicide. *Dis. Nerv. Syst.* **1972**, *33*, 546–556.
79. Schrauzer, G.N.; Shrestha, K.P. Lithium in drinking water and the incidences of crimes, suicides, and arrests related to drug addictions. *Biol. Trace Elem. Res.* **1990**, *25*, 105–113. [CrossRef] [PubMed]
80. Schrauzer, G.N. Lithium: Occurrence, dietary intakes, nutritional essentiality. *J. Am. Coll. Nutr.* **2002**, *21*, 14–21. [CrossRef] [PubMed]
81. Ohgami, H.; Terao, T.; Shiotsuki, I.; Ishii, N.; Iwata, N. Lithium levels in drinking water and risk of suicide. *Br. J. Psychiatry* **2009**, *194*, 464–465. [CrossRef]
82. Kabacs, N.; Memon, A.; Obinwa, T.; Stochl, J.; Perez, J. Lithium in drinking water and suicide rates across the East of England. *Br. J. Psychiatry* **2011**, *198*, 40–407. [CrossRef]
83. Kapusta, N.D.; Mossaheb, N.; Etzersdorfer, E.; Hlavin, G.; Thau, K.; Willeit, M.; Praschak-Rieder, N.; Sonneck, G.; Leithner-Dziubas, K. Lithium in drinking water and suicide mortality. *Br. J. Psychiatry* **2011**, *198*, 346–350. [CrossRef] [PubMed]
84. Blüml, V.; Regier, M.D.; Hlavin, G.; Rockett, I.R.; König, F.; Vyssoki, B.; Bschor, T.; Kapusta, N.D. Lithium in the public water supply and suicide mortality in Texas. *J. Psychiatr. Res.* **2013**, *47*, 407–411. [CrossRef]
85. Giotakos, O.; Nisianakis, P.; Tsouvelas, G.; Giakalou, V.V. Lithium in the public water supply and suicide mortality in Greece. *Biol. Trace Elem. Res.* **2013**, *156*, 376–379. [CrossRef] [PubMed]
86. Sugawara, N.; Yasui-Furukori, N.; Ishii, N.; Iwata, N.; Terao, T. Lithium in tap water and suicide mortality in Japan. *Int. J. Environ. Res. Public Health* **2013**, *10*, 6044–6048. [CrossRef]
87. Huber, R.S.; Kim, N.; Renshaw, C.E.; Renshaw, P.F.; Kondo, D.G. Relationship between altitude and lithium in groundwater in the United States of America: Results of a 1992–2003 study. *Geospat Health* **2014**, *9*, 231–235. [CrossRef] [PubMed]
88. Giotakos, O.; Tsouvelas, G.; Nisianakis, P.; Giakalou, V.; Lavdas, A.; Tsiamitas, C.; Panagiotis, K.; Kontaxakis, V. A negative association between lithium in drinking water and the incidence of homicides, in Greece. *Biol. Trace Elem. Res.* **2015**, *164*, 165–168. [CrossRef] [PubMed]
89. Ishii, N.; Terao, T.; Araki, Y.; Kohno, K.; Mizokami, Y.; Shiotsuki, I.; Hatano, K.; Makino, M.; Kodama, K.; Iwata, N. Low risk of male suicide and lithium in drinking water. *J. Clin. Psychiatry* **2015**, *76*, 319–326. [CrossRef] [PubMed]
90. Pompili, M.; Vichi, M.; Dinelli, E.; Pycha, R.; Valera, P.; Albanese, S.; Lima, A.; De Vivo, B.; Cicchella, D.; Fiorillo, A.; et al. Relationships of local lithium concentrations in drinking water to regional suicide rates in Italy. *World J. Biol. Psychiatry* **2015**, *16*, 567–574. [CrossRef]
91. Vita, A.; De Peri, L.; Sacchetti, E. Lithium in drinking water and suicide prevention: A review of the evidence. *Int. Clin. Psychopharmacol.* **2015**, *30*, 1–5. [CrossRef] [PubMed]
92. Goldstein, M.R.; Mascitelli, L. Is violence in part a lithium deficiency state? *Med. Hypotheses* **2016**, *89*, 40–42. [CrossRef]
93. Shiotsuki, I.; Terao, T.; Ishii, N.; Takeuchi, S.; Kuroda, Y.; Kohno, K.; Mizokami, Y.; Hatano, K.; Tanabe, S.; Kanehisa, M.; et al. Trace lithium is inversely associated with male suicide after adjustment of climatic factors. *J. Affect. Disord.* **2016**, *189*, 282–286. [CrossRef] [PubMed]
94. Ando, S.; Koike, S.; Shimodera, S.; Fujito, R.; Sawada, K.; Terao, T.; Furukawa, T.A.; Sasaki, T.; Inoue, S.; Asukai, N.; et al. Lithium levels in tap water and the mental health problems of adolescents: An individual-level cross-sectional survey. *J. Clin. Psychiatry* **2017**, *78*, e252–e256. [CrossRef]
95. Awad, A.; Cipriani, A. Prophylactic mood stabilization: What is the evidence for lithium exposure in drinking water? *Bipolar Disord.* **2017**, *19*, 601–602. [CrossRef] [PubMed]
96. Kanehisa, M.; Terao, T.; Shiotsuki, I.; Kurosawa, K.; Takenaka, R.; Sakamoto, T.; Shigemitsu, O.; Ishii, N.; Hatano, K.; Hirakawa, H. Serum lithium levels and suicide attempts: A case-controlled comparison in lithium therapy-naive individuals. *Psychopharmacology* **2017**, *234*, 3335–3342. [CrossRef]
97. Knudsen, N.N.; Schullehner, J.; Hansen, B.; Jørgensen, L.F.; Kristiansen, S.M.; Voutchkova, D.D.; Gerds, T.A.; Andersen, P.K.; Bihrmann, K.; Grønbæk, M.; et al. Lithium in drinking water and incidence of suicide: A nationwide individual-level cohort study with 22 years of follow-up. *Int. J. Environ. Res. Public Health* **2017**, *14*, 627. [CrossRef] [PubMed]
98. König, D.; Baumgartner, J.; Blüml, V.; Heerlein, A.; Téllez, C.; Baus, N.; Kapusta, N.D. Impact of natural lithium resources on suicide mortality in Chile 2000–2009, a geographical analysis. *Neuropsychiatrie* **2017**, *31*, 70–76. (In German) [CrossRef] [PubMed]

99. Brown, E.E.; Gerretsen, P.; Pollock, B.; Graff-Guerrero, A. Psychiatric benefits of lithium in water supplies may be due to protection from the neurotoxicity of lead exposure. *Med. Hypotheses* **2018**, *115*, 94–102. [CrossRef]
100. Fajardo, V.A.; LeBlanc, P.J.; Fajardo, V.A. Trace lithium in Texas tap water is negatively associated with all-cause mortality and premature death. *Appl. Physiol. Nutr. Metab.* **2018**, *43*, 412–414. [CrossRef]
101. Kurosawa, K.; Terao, T.; Kanehisa, M.; Shiotsuki, I.; Ishii, N.; Takenaka, R.; Sakamoto, T.; Matsukawa, T.; Yokoyama, K.; Ando, S.; et al. Naturally absorbed polyunsaturated fatty acids, lithium, and suicide-related behaviors: A case-controlled study. *J. Affect. Disord.* **2018**, *241*, 200–205. [CrossRef] [PubMed]
102. Oliveira, P.; Zagalo, J.; Madeira, N.; Neves, O. Lithium in Public Drinking Water and Suicide Mortality in Portugal: Initial Approach. *Acta Med. Port.* **2019**, *32*, 47–52. [CrossRef]
103. Palmer, A.; Cates, M.E.; Gorman, G. Association between lithium in drinking water and incidence of suicide across 15 Alabama counties. *Crisis* **2019**, *40*, 93–99. [CrossRef] [PubMed]
104. Barjasteh-Askari, F.; Davoudi, M.; Amini, H.; Ghorbani, M.; Yaseri, M.; Yunesian, M.; Mahvi, A.M.; Lester, D. Relationship between suicide mortality and lithium in drinking water: Systematic review and meta-analysis. *J. Affect. Disord.* **2020**, *264*, 234–241. [CrossRef]
105. Del Matto, L.; Muscas, M.; Murru, A.; Verdolini, N.; Anmella, G.; Fico, G.; Corponi, F.; Carvalho, A.F.; Samalin, L.; Carpiniello, B.; et al. Lithium and suicide prevention in mood disorders and in the general population: A systematic review. *Neurosci. Biobehav. Rev.* **2020**, *116*, 142–153. [CrossRef] [PubMed]
106. Eyre-Watt, B.; Mahendran, E.; Suetani, S.; Firth, J.; Kisely, S.; Siskind, D. The association between lithium in drinking water and neuropsychiatric outcomes: A systematic review and meta-analysis from across 2678 regions containing 113 million people. *Aust. N. Z. J. Psychiatry* **2021**, *55*, 139–152. [CrossRef] [PubMed]
107. Kozaka, N.; Takeuchi, S.; Ishii, N.; Terao, T.; Kuroda, Y. Association between lithium in tap water and suicide mortality rates in Miyazaki Prefecture. *Environ. Health Prev. Med.* **2020**, *25*, 26–32. [CrossRef]
108. Memon, A.; Rogers, I.; Fitzsimmons, S.M.D.D.; Carter, B.; Strawbridge, R.; Hidalgo-Mazzei, D.; Young, A.H. Association between naturally occurring lithium in drinking water and suicide rates: Systematic review and meta-analysis of ecological studies. *Br. J. Psychiatry* **2020**, *217*, 667–678. [CrossRef] [PubMed]
109. Kugimiya, T.; Ishii, N.; Kohno, K.; Kanehisa, M.; Hatano, K.; Hirakawa, H.; Terao, T. Lithium in drinking water and suicide prevention: The largest nationwide epidemiological study from Japan. *Bipolar Disord.* **2021**, *23*, 33–40. [CrossRef] [PubMed]
110. Liaugaudaite, V.; Naginiene, R.; Raskauskiene, N.; Mickuviene, N.; Bunevicius, A.; Sher, L. Relationship between lithium levels in drinking water and suicide rates: A nationwide study in Lithuania. *Arch. Suicide Res.* **2021**, *25*, 340–352. [CrossRef] [PubMed]
111. Araya, P.; Martínez, C.; Barros, J. Lithium in drinking water as a public policy for suicide prevention: Relevance and considerations. *Front. Public Health* **2022**, *10*, 805774. [CrossRef]
112. Izsak, B.; Hidvegi, A.; Balint, L.; Malnasi, T.; Vargha, M.; Pandics, T.; Rihmer, Z.; Dome, P. Investigation of the association between lithium levels in drinking water and suicide mortality in Hungary. *J. Affect. Disord.* **2022**, *298*, 540–547. [CrossRef]
113. Kawada, T. Lithium in drinking water and suicide risk: Review. *EXCLI J.* **2022**, *21*, 571–572.
114. Liaugaudaite, V.; Raskauskiene, N.; Naginiene, R.; Mickuviene, N.; Sher, L. Association between lithium levels in drinking water and suicide rates: Role of affective disorders. *J. Affect. Disord.* **2022**, *298*, 516–521. [CrossRef]
115. Fadaei, A. An investigation into the association between suicide mortality rate and lithium levels in potable water: A review study. *Int. Clin. Psychopharmacol.* **2023**, *38*, 73–80. [CrossRef] [PubMed]
116. Rihmer, Z.; Dome, P. Fortification of drinking water with lithium to reduce suicide in the population. *Eur. Neuropsychopharmacol.* **2023**, *71*, 3–5. [CrossRef]
117. Bjørklund, G.; Storchylo, O. Lithium in drinking water and its potential impact on mental health: A review. *Curr. Med. Chem.* **2024**. *online ahead of print*. [CrossRef] [PubMed]
118. Harandi, H.; Ahmadinia, H.; Ghaffarian-Bahraman, A.; Alimoradi, H.; Nasab, Z.B.; Rezaeian, M. Correlation between lithium concentrations in drinking water and suicide attempt in the southeast of Iran. *Environ. Monit. Assess.* **2024**, *196*, 1144. [CrossRef] [PubMed]
119. Pichler, E.M.; Fartacek, C.; Miller-Doebeling, B.; Walter, M.; Plöderl, M. Too early to add lithium to drinking water? No association between lithium and suicides in a pre-registered Swiss study. *J. Affect. Disord.* **2024**, *367*, 598–605. [CrossRef]
120. Terao, T.; Hirakawa, H.; Muronaga, M.; Izumi, T.; Kohno, K. Trace Lithium for Suicide Prevention and Dementia Prevention: A Qualitative Review. *Pharmaceuticals* **2024**, *17*, 1486. [CrossRef]

pharmaceuticals

MDPI

Review

An Overview of the Effects of Lithium on Alzheimer's Disease: A Historical Perspective

Marcia Radanovic [1,2,†], Monique Patricio Singulani [1,2,3,†], Vanessa de Jesus R. De Paula [1,2], Leda Leme Talib [1,2] and Orestes Vicente Forlenza [1,2,3,*]

[1] Laboratory of Neuroscience (LIM-27), Department and Institute of Psychiatry, Hospital das Clínicas da Faculdade de Medicina da Universidade de São Paulo (HCFMUSP), São Paulo 05403-010, SP, Brazil; marciaradanovic@gmail.com (M.R.); moniquesingulani@gmail.com (M.P.S.); vanessajrpaula@gmail.com (V.d.J.R.D.P.); ledatalib@gmail.com (L.L.T.)

[2] Instituto Nacional de Biomarcadores em Neuropsiquiatria (INBioN), Conselho Nacional de Desenvolvimento Científico e Tecnológico, São Paulo 05403-010, SP, Brazil

[3] Centro de Neurociências Translacionais (CNT), Faculdade de Medicina da Universidade de São Paulo (FMUSP), São Paulo 05403-010, SP, Brazil

* Correspondence: forlenza@usp.br

† These authors contributed equally to this work.

Abstract: Lithium was introduced into psychiatric practice in the late nineteenth century and has since become a standard treatment for severe psychiatric disorders, particularly those characterized by psychotic agitation. It remains the most effective agent for managing acute mania and preventing relapses in bipolar disorder. Despite potential adverse effects, lithium's use should be carefully considered relative to other treatment options, as these alternatives may present distinct safety and tolerability profiles. The World Health Organization classifies lithium salts as 'essential' medications for inclusion in global healthcare systems. Over the past two decades, the growing recognition of lithium's efficacy—extending beyond mood stabilization to include reducing suicide risk and inducing neuroprotection—has led to its incorporation into clinical practice guidelines. Current research, particularly from translational models, suggests that lithium's pleiotropic effects benefit not only mental and brain health but also other organs and systems. This supports its potential as a therapeutic candidate for neurological conditions, particularly those associated with neurodegenerative processes. This article will discuss the historical background, discovery, and early experimentation of lithium in psychiatry. We will also review its mechanisms of action and discuss its potential in the treatment and prevention of neurodegenerative disorders, focusing on Alzheimer's disease.

Keywords: lithium salts; lithium therapy; neurodegenerative diseases; Alzheimer's disease; neuroprotection

1. Introduction

Lithium is a soft, silvery-white alkali metal with an atomic mass of 6.941 u and an atomic number of 3 [1]. Its therapeutic potential has been recognized for millennia, with the first recorded use dating back over two thousand years. Ancient civilizations discovered the medicinal benefits of lithium-rich waters, which were believed to have healing properties for a range of physical and mental health conditions [2]. One of the earliest documented references occurs in the early second century C.E., when the Greek Soranus of Ephesus, a renowned physician expert of the time, recommended the use of alkaline waters to promote mental health. He recommended, particularly for individuals suffering from conditions such as mania and melancholia, to immerse themselves in these mineral-rich springs as a treatment form. Later, this practice led to the eventual identification of a link between lithium-containing mineral springs and the management of manic episodes. By the 19th century, the therapeutic use of water from various hydro-mineral resorts, including the famous Eau de Vichy, became widely popular. These mineral waters were believed to provide relief for a broad spectrum of both physical and mental health disorders, including mania, melancholia, epilepsy, arthritis, gout, and even cancer [2–5].

The widespread adoption of lithium marked the beginning of modern psychopharmacology and continues to be the most well-established mood stabilizer for bipolar disorder (BD) [6]. Moreover, substantial evidence demonstrates its efficacy in preventing suicide and treating neurodegenerative diseases, further solidifying its role as a potential therapeutic agent in psychiatry.

2. The Journey of Lithium: From Discovery to Clinical Applications

The discovery of lithium can be traced back to the mid-18th century and is credited to José Bonifácio de Andrada e Silva, a Brazilian statesman who played a key role in his country's struggle for independence from Portugal. In addition to his political accomplishments, Andrada e Silva was also a scientist, and his notable contribution as a mineralogist was the identification of petalite crystals on the Swedish island of Utö, renowned for its rich iron mines, from which lithium was later extracted [7]. During this period, Reverend Edward Daniel Clarke, a professor of mineralogy at the University of Cambridge, was among the researchers who obtained access to petalite samples and published a detailed description of its constituents; however, a small fraction remained unidentified [2,5]. In 1817, while analyzing a petalite specimen in the laboratory of Jöns Jacob Berzelius, one of the founders of modern chemistry, the 25-year-old Swedish chemist Johan August Arfwedson identified a previously unknown element. Subsequently, Berzelius named the new element 'lithium', derived from the Greek word 'lithos' (stone), to reflect its mineral origin [2,4].

William Thomas Brande, an English chemist and geologist, became the first to isolate lithium in its pure form in 1821 by performing electrolysis on lithium oxide. It was not until 1855 that the British chemist Augustus Matthiessen and the German chemist Robert Bunsen developed a process that enabled the large-scale production of lithium through the electrolysis of lithium chloride. This discovery not only facilitated the systematic investigation of lithium's properties across various scientific and industrial domains but also marked a pivotal turning point in the initiation of its widespread commercialization for various applications. Furthermore, with the availability of sufficient quantities, it was incorporated into medical treatments [2,4,5].

The early medical use of lithium began with its administration for the treatment of gout, and later expanded to include conditions related to the bladder, kidneys, and gallstones. In 1843, Alexander Ure, a Scottish surgeon, proposed using lithium carbonate to treat renal stones, based on his observation that it could dissolve uric acid crystals. In 1860, Ure suggested treating patients with urinary stones by injecting lithium salts into their urinary bladders, but the treatment was not clinically successful. His findings, on the other hand, contributed to the belief that lithium could also be beneficial in the treatment of gout, thus enabling continued research on its effects [2,5,8].

Later, in 1847, Alfred Baring Garrod, a London physician, detected elevated concentrations of uric acid in the blood of patients suffering from gout. He began investigating the use of lithium salts as a treatment for dissolving uric acid crystals. His research culminated in the publication of 'The Nature and Treatment of Gout and Rheumatic Gout' in 1859, a work that significantly contributed to the broader understanding and application of lithium in medical treatments [9]. It was also noted in that same decade that a significant number of individuals diagnosed with gout exhibited symptoms of depression. Garrod hypothesized that the effects of excessive uric acid could impact the central nervous system, resulting in what he termed 'brain gout', a condition he associated with depression, based on his empirical observations. He later recommended the use of lithium to treat mental illness [9]. However, it was the writings of Alexander Haig (1853–1924) on 'uric acid diathesis' that ultimately contributed to the widespread use of lithium. Haig, in agreement with Garrod's hypothesis, suggested that excess uric acid was responsible for a wide range of ailments, including neuropsychiatric disorders, and recommended the use of lithium salts to treat these conditions [10].

The use of lithium in psychiatry dates back to the 1870s, when American surgeon general William Hammond, a professor at the Bellevue Hospital Medical College, presented evidence of effectively treating acute manic patients with lithium bromide. However, he was unable to determine if the outcomes were attributable to lithium or bromide [11]. In the 1880s and 1890s, Danish psychiatrists Carl Georg Lange and Frederick Lange were the first to systematically use lithium for both acute and prophylactic treatment of melancholic depression. The Lange brothers reported the successful treatment of approximately 2000 depressed outpatients with lithium carbonate over a 20-year period [12]. Carl also noted that some patients with depression treated with lithium carbonate showed elevated levels of urinary sediment, which he attributed to the liberation of uric acid. Even though the uric acid hypothesis was eventually refuted, lithium treatment provided tangible benefits in various general medical and neuropsychiatric conditions [12]. However, ongoing debates regarding the validity of the hypothesis in relation to neuropsychiatric disorders and reports of severe intoxications and fatalities contributed to the gradual decline in the use of lithium [13].

3. The Modern Revival of Lithium

Interest in lithium reemerged during the early 19th century when David Culbreth, a pharmacology professor at the University of Maryland, described lithium bromide as 'the most hypnotic of all bromides' and advocated its use in the treatment of epilepsy. Despite this, lithium salts were not integrated into the treatment of mental illnesses until the late 1940s [14].

The foundational research that contributed to the advancement and dissemination of the use of lithium salts as a successful treatment for manic-depressive psychoses and BD is credited to Australian psychiatrist John F. J. Cade. Based on his interventional studies, Cade demonstrated the antimanic effects of lithium, describing it as 'lithium salts might well be an important addition to the anti-convulsant armamentarium', and detailing his investigation of the 'protective effect' of lithium [15]. His seminal paper was published in the 3 September 1949 issue of the *Medical Journal of Australia*, under the title 'Lithium Salts in the Treatment of Psychotic Excitement' [16]. Cade hypothesized that the etiology of 'manic-depressive insanity' was due to a dysregulation of the body's metabolism. He proposed that mania resulted from a 'state of intoxication of a normal product of the body circulating in excess', whereas melancholia was a 'corresponding deprivative condition'. To investigate these metabolites—the presumed 'toxic agent'—Cade conducted what he described as an 'extraordinarily crude differential toxicity test' to identify substances excreted in excess in the urine of patients with psychiatric disorders, which could contribute to distinguishing between those with mania and those with depression based on their symptoms [16].

Cade used a scientific approach that involved injecting samples of concentrated urine from manic, melancholic, and schizophrenic patients, as well as from healthy people, into the peritoneal cavity of guinea pigs. Although the mode of death in the animals continued to be consistent, Cade's experiment indicated that the urine of human subjects diagnosed with mania was harmful to animals compared to urine from other individuals. Initially, Cade hypothesized that urea might be the 'toxic agent', as intraperitoneal injections of this urine constituent caused the same mode of death in the guinea pigs. However, during his analysis, Cade noted that, despite similar urea levels, the urine of manic patients was significantly more toxic than that of non-manic individuals. He suggested that another substance might be contributing to the increased toxicity of urea after this finding. At that time, Cade had already noticed that uric acid seemed to increase the toxicity of urea. However, due to the water-insoluble nature of uric acid, lithium urate, the most soluble urate, was used as an alternative. Cade observed that the toxicity was significantly lower than expected, a phenomenon he described as the 'great paradox'. He speculated that the lithium-ion might be exerting a protective effect. To further explore the findings of his previous experiment, Cade used a urea solution with lithium carbonate instead of lithium urate. He observed a reduction in lethality in the guinea pigs, which confirmed that lithium itself had a protective effect against the toxic impact of urea [16]. Cade also observed that when lithium salts were tested alone (i.e., without urea), they caused anxiolytic effects in guinea pigs [15,16].

Cade's unexpected finding led him to shift his focus toward investigating the tranquilizing effects of lithium on his patients. To establish a safe dosage of lithium salts for human use, he first ingested them himself. Following this, he conducted a pilot trial involving patients diagnosed with mania, melancholia, and early dementia, which led to the hypothesis that lithium-ion deficiency might contribute to these conditions [17,18]. Cade treated 10 manic patients, 6 schizophrenic patients, and 3 melancholic patients with lithium, specifically lithium citrate and lithium carbonate. All patients exhibited recovery within a few days to weeks; however, relapse was observed upon discontinuation of lithium or in cases of non-compliance. While the psychotic symptoms of schizophrenic patients remained unchanged, a notable reduction in restlessness and agitation was observed, with three patients becoming calm and responsive for the first time in years. However, despite lithium treatment, melancholic patients continued to experience distress [17,18].

Although Cade's interventional studies lacked standardized measurement methods, statistical rigor, and other methodological elements characteristic of contemporary research, his findings significantly influenced psychiatric practice [17,18]. Despite these promising results, the research faced challenges due to lithium's side effects, which, in some cases, led to toxicity and even fatalities. Two years later, Cade's first patient (Case I, W.B.) died of lithium poisoning due to the prescribed doses, prompting him to discontinue his research and cease prescribing lithium [13,15,17].

4. The Legacy of Lithium: From Therapeutic Validation to Approval

Although lithium treatment was considered controversial, primarily due to the occasional occurrence of severe toxic complications, the research on lithium remained ongoing throughout the early 1950s [2,5].

The first clinical report in the global literature following Cade's publication appeared in 1951 in the *Medical Journal of Australia*. The study, entitled 'The Lithium Treatment of Maniacal Psychosis', was accomplished by Edward M. Trautner at the Department of Physiology at the University of Melbourne and Charles H. Noack at Mont Park Mental Hospital, Melbourne [19]. They conducted an open trial involving over 100 psychiatric patients with various diagnoses, who were treated with 0.6 g of lithium carbonate or 1.2 g of lithium citrate for durations ranging from two weeks to over one year. As a result, in most cases, Trautner et al. reported that 'lithium treatment was found to be very beneficial in terminating and preventing the maniacal phase in cases of mania, hypomania and recurrent mania', and added that 'no cumulative effects or incidents of grave intoxication were observed in patients'. A significant aspect of this study was the introduction of blood lithium level assessment in all patients through flame spectrophotometry assays. With its approach, Trautner emphasized the importance of determining lithium levels for both safety monitoring and the validation of optimal therapeutic plasma concentrations [19].

In subsequent years, Trautner's findings were confirmed by other studies; however, considerable skepticism remained regarding the long-term use of lithium, primarily due to concerns about the relationship between appropriate dosage, plasma lithium levels, and toxicity, which were widely recognized as unresolved issues at the time [20]. In 1995, Trautner presented notable findings on the pharmacokinetic properties of lithium, focusing on its effects on ionic balance, retention, excretion, and toxicity. These observations were performed on patients receiving lithium salts during the manic phase compared with their return to euthymia, as well as on healthy individuals for comparison. In his study published in *The Medical Journal of Australia* and titled 'The Excretion and Retention of Ingested Lithium and Its Effect on the Ionic Balance of Man', Trautner et al. reported that 90% to 95% of ingested lithium was excreted in urine, with most eliminated within 6–8 h and the rest over 2 weeks [21]. Lithium ingestion caused disturbances in water and ionic balance, particularly sodium (Na^+), for up to 2 days. With prolonged moderate intake, a stable state was reached where lithium excretion matched intake, and ionic imbalances decreased. In this stable state, lithium retention equaled one and a half daily doses, and plasma levels correlated with the dose. However, in patients on high doses, stability was not maintained, and clinical instability, such as low plasma Na^+ levels and susceptibility to toxic complications, was observed. Trautner also identified early symptoms of lithium poisoning, with a danger level at 3 mEq/L of plasma and a lethal level at 4–5 mEq/L [21].

The understanding of lithium's therapeutic effects further progressed through investigations by Danish psychiatrist Mogens Schou and other researchers, who conducted a series of studies under more rigorous conditions. This led to a randomized controlled clinical trial on lithium treatment in mania patients, published in 1954, which demonstrated its efficacy for the majority of individuals with BD [22]. In 1967, Danish psychiatrist Poul C. Baastrup, along with Schou, demonstrated that long-term lithium treatment could not only alleviate manic episodes but also prevent depressive recurrences [23]. Confirming these findings, Baastrup and his collaborators published a 1970 study in *The Lancet* titled 'Prophylactic Lithium: Double-Blind Discontinuation in Manic-Depressive and Recurrent-Depressive Disorders'. This double-blind discontinuation clinical trial included fifty patients with manic-depressive illness and thirty-four with recurrent endogenous depression, all of whom had been receiving open lithium treatment. The results demonstrated that lithium carbonate exerted a prophylactic effect in preventing further recurrences of endogenous affective disorders [24], thereby reinforcing its therapeutic efficacy and prophylactic potential [25].

While lithium gained recognition in several countries during the 1950s [19,22,26], interest in its use in the United States did not grow until a decade later. In 1960, Australian psychiatrist Sam Gershon, an early pioneer of lithium research in Melbourne, collaborated with Trautner to conduct a major clinical trial in the United States during the 1960s [27]. This study represents the first report of lithium treatment in North America, with seventeen patients diagnosed with manic excitement being treated at the Royal Victoria Hospital in Montreal. As a result, all patients, except for one, responded positively to lithium treatment. The authors also showed that lithium dosages ranging from 1.2 to 5.4 g, for which no poisoning occurred, resulted in maximal plasma levels of 0.8–2.8 mEq/L and highlighted lithium's specificity for the treatment of manic excitement [27,28]. They also concluded that there were no significant barriers to the widespread use of lithium, which is one of the few available targeted therapies in modern psychiatry [29].

In fact, lithium was banned from medical practice in the US until the 1970s because of its unwanted effects. However, in 1968, the clinical significance of lithium was recognized in a special section of the *American Journal of Psychiatry*. In the early 1970s, the U.S. Food and Drug Administration (FDA) approved the use of lithium for the treatment of mania, and in 1974, it was approved for maintenance therapy of patients with mania, making the United States the 50th country to introduce it to the market [5,29]. Lithium had also been approved for medical use in several other countries, including France in 1961, the United Kingdom in 1966, Germany in 1967, and Italy in 1970 [4]. In retrospect, as evidence of lithium's usefulness increased, strong support from the pharmaceutical industry might have been anticipated. However, since lithium is a naturally occurring element, making it an inexpensive, non-patentable drug, no American pharmaceutical company was willing to manufacture it, even after FDA approval, due to its lack of profitability. As a result, lithium research in the late 19th century progressed slowly, particularly in determining its safety limits for clinical use [3,6,28].

5. Broadening the Use of Lithium in Neurocognitive Functions

Since its FDA approval, further research has substantiated lithium's therapeutic and prophylactic efficacy in managing acute mania and preventing bipolar disorder, while its application has extended to other medical fields. However, long-term administration has been linked to adverse effects and complications, including memory impairment and kidney dysfunction. This caught the attention of the medical community, raising concerns about the potential for lithium to induce irreversible memory disorders in patients with psychiatric illnesses [30]. However, a double-blind crossover clinical trial conducted in patients with affective disorders demonstrated that lithium carbonate treatment (0.5–2.2 mEq/L) led to a slowdown in performance on certain perceptual-motor tests but did not impair memory function [31]. In contrast, another study found that bipolar patients treated with lithium carbonate recalled significantly fewer words compared to unmedicated patients, suggesting a potential impact on verbal memory [32]. Further comparisons with patients diagnosed with affective disorders and treated with tricyclic antidepressants also indicated that the lithium-treated groups exhibited memory impairment [33]. Presumably, the adverse effects of lithium on cognition are associated with the use of higher doses. Furthermore, it is important to determine whether there is conclusive evidence of such effects, considering the psychiatric status of lithium-treated patients and whether the battery of memory and cognitive tests used was the most appropriate for each case [34–36].

The positive effects of lithium treatment on neurobiological and neurocognitive functions began to be demonstrated in the late 1970s. Physicians Kenneth H. Williams and Gerald Goldstein were the first to suggest, in a paper published in the *American Journal of Psychiatry* in 1979, that lithium could play a role in treating dementia-like symptoms. Their study, titled 'Cognitive and Affective Responses to Lithium in Patients with Organic Brain Syndrome', examined patients with a combination of dementia and agitated depression who were hospitalized for organic brain syndrome. These patients received lithium carbonate at doses of 900–1200 mg/day. Of the 10 cases they evaluated, 8 were described as having shown 'improved dramatically in affect and cognition with lithium after having shown no significant change with other psychotropic medications' [37]. In 1990, Calil et al. [38] showed that healthy individuals taking lithium, with serum levels ranging from 0.6 to 1.0 mEq/L, exhibited slower reaction times but demonstrated improved performance on cognitive tasks. In the same decade, Clarke et al. [39] studied the effect of lithium on cerebrospinal fluid (CSF) amyloid-beta (Aβ) precursor protein (APP)—a transmembrane protein that plays a central role in the pathophysiology of Alzheimer's disease (AD)—in dementia patients. Their findings showed that the mean APP695 value, the most common APP isoform in the brain and a key player in Aβ production—a hallmark of AD, was lower in demented patients who had received lithium salts compared to those treated with other medications, such as antidepressants, antipsychotics, and tranquilizers. This study sparked interest in lithium's neuroprotective properties for AD treatment, as lithium appeared to selectively modulate APP, may influence the disease's pathological processes. But the interest in lithium's therapeutic action grew stronger when Paul Grof (2010) introduced the concept of 'excellent lithium responders', a specific subgroup of BD patients in whom lithium monotherapy can restore normality in clinical profile and cognitive function, compared to age-matched, healthy control subjects [40]. A few years later, considerable evidence accumulated showing the neuroprotective effect of lithium at therapeutic and subtherapeutic doses [6,36,41–43].

6. Lithium and Its Impact on Human Health

Natural lithium is present in human nutrition through food intake, predominantly in plant-based foods and drinking water. Traces of lithium have been detected in the human body, with an average concentration of approximately 7 mg [44]. Lithium is an essential trace element involved in various cellular processes, influencing the functions of multiple enzymes, hormones, and vitamins. The fact that lithium is a monovalent cation with chemical properties similar to potassium (K^+) and Na^+ suggests that it plays a key role in maintaining brain homeostasis. Lithium is equally distributed between intra- and extracellular compartments and enters cells mainly through voltage-gated Na^+ channels, with its influx exceeding its efflux. Since lithium enters cells through electrical signaling, it accumulates more selectively in neurons with higher synaptic activity. As a result, lithium is more likely to interact with neurotransmitters and receptors in the brain, modulating neuronal plasticity, membrane components, transport mechanisms, and intracellular signaling molecules. At a structural level, it is via such a mechanism that lithium appears to preserve or increase the volume of brain structures involved in emotional regulation, such as the prefrontal cortex, hippocampus, and amygdala, possibly reflecting its neuroprotective effects [45,46]. This is supported by evidence of post-mortem studies that have indicated that the cerebellum, cerebrum, and kidneys retain more lithium than other organs, suggesting its potential influence on signaling pathways in key brain regions [1,44,45,47].

In addition to lithium, the other alkali metals, including Na^+, K^+, rubidium (Rb), cesium (Cs), and francium (Fr), have some function in the organism. In the human body, only Na^+ and K^+ are essential and vital to life, playing crucial roles in various physiological processes, such as the conduction of electrical impulses in excitable cells. Even a slight variation in their concentrations, either in the intra- or extracellular compartments, can lead to acute dysfunction of vital organs such as the brain, heart, and kidneys, potentially resulting in a life-threatening state. In contrast, Rb and Cs are considered non-essential elements for humans and have no known biological role [45,48].

According to nutritional studies, dietary lithium intake is necessary, provided it does not exceed 1 mg/day for a 70 kg adult (14.3 μg/kg body weight) [44,47]. Low-level exposure to naturally occurring lithium in the environment, such as in drinking water and certain food, including vegetables, fruits, and grains, e.g., cauliflower, tea plants, coriander leaves, tomatoes, garlic, nutmeg, cumin seeds, onions, green chilies, rice, and wheat, has been shown to benefit mental health [44]. Although natural lithium intake is significantly lower than therapeutic doses, studies have documented that low concentrations in drinking water consumed daily are associated with increased human lifespan. This effect is attributed to lithium's anti-suicidal properties, its ability to prevent depressive states, and its role in reducing the incidence of dementia; however, the underlying neurobiological mechanisms are not yet fully understood [3,36,49].

Another important characteristic of lithium is its ability to easily distribute throughout the body via both intra- and extracellular fluids, due to its free protein binding. Lithium can cross the placenta to reach the developing fetus, and is also secreted in breast milk. As a result, lithium concentrations in both mothers and fetuses attain equilibrium, suggesting that it may contribute to neurodevelopmental health. In addition, during breastfeeding, the lithium concentration in milk decreases to approximately half of the maternal serum levels [46,50]. The control of lithium levels is particularly challenging during pregnancy due to normal physiological changes, especially in renal function. Increased glomerular filtration rate during pregnancy leads to decreased blood lithium concentrations and an increased risk of relapse in women with mania. As a result, clinicians often increase the lithium dose during pregnancy. However, later in pregnancy and during the early postpartum period, when glomerular filtration rate returns to preconception levels, the increased dosage may lead to toxic blood lithium levels [50]. Wesseloo et al. [50] showed that lithium initiated during pregnancy, either as treatment for an episode or as an alternative to a mood stabilizer, reduced the risk of teratogenicity when started before conception and continued throughout pregnancy and the postpartum period. The lowest blood lithium levels occurred during the second trimester; in the third trimester and postpartum period, lithium levels progressively returned to preconception levels, highlighting the importance of controlling lithium during this critical gestational period [50].

7. Managing Lithium: Therapeutic Index and Risk to Toxicity

Lithium was the first specific psychotropic agent, with sedatives such as bromides and paraldehyde, being its predecessors. Currently, it remains the benchmark for the treatment of bipolar disorder and is considered the standard reference among mood stabilizers. Although there are other approved medications for this condition, the use of lithium must be carefully assessed in comparison to alternative treatments, given that these alternatives may have differing safety and tolerability profiles. Nonetheless, the management of lithium treatment is challenging due to its narrow therapeutic index and the impact on kidney function, both of which heighten the risk of toxicity [51].

Meticulous attention to dosing, monitoring, and titration is essential. Lithium is prescribed for managing acute manic and mixed episodes, as well as for long-term maintenance treatment in patients aged 12 or older. The therapeutic dose typically ranges from 300 to 2700 mg/day, with the optimal steady-state serum concentration of lithium falling between 0.7 and 1.2 mEq/L. In elderly individuals, lower doses of lithium are required, with a mean dose just above 400 mg/day, to achieve the desired serum concentration of approximately 0.5 mmol/L. For children over the age of 12, serum lithium concentrations should generally align with those observed in younger adults, ranging between 0.6 and 1.2 mmol/L. Serum concentrations of lithium exceeding 1.5 mEq/L may lead to symptoms of mild intoxication, including fatigue, tremor, nausea, diarrhea, blurred vision, vertigo, confusion, and increased deep tendon reflexes. Levels surpassing 2.5 mEq/L may cause more severe neurological complications, such as seizures, coma, cardiac dysrhythmia, and permanent neurological impairment, often affecting the cerebellum. Careful management is essential when using lithium, and it should be avoided in patients with a history of acute myocardial infarction, acute kidney failure, or certain rare heart rhythm disorders [51,52].

8. Advancing Lithium Research Through Translational Studies

Despite its extensive clinical use for over eight decades, uncertainties remain regarding the mechanisms underlying lithium's effects on mood and behavior. Lithium is often labeled a 'dirty' drug due to its interaction with various intracellular signaling pathways, including Wnt/β-catenin, adenylate cyclase, telomerase, bisphosphate nucleotidase, β-arrestin, and cyclooxygenase. Additionally, it affects the actions of GABA (gamma-aminobutyric acid) and NMDA (*N*-methyl-D-aspartate) receptors, and contributes to calcium homeostasis [53,54].

The primary molecular targets of lithium are the enzymes inositol monophosphatase (IMPase) and glycogen synthase kinase-3 beta (GSK-3β). These enzymes play critical roles in various cell types, directly or indirectly influencing metabolic processes such as apoptosis, autophagy, cytoskeletal remodeling, gene regulation, the cell cycle, neurotrophic support, energy metabolism, mitochondrial function, oxidative stress, and the inflammatory response [49].

Lithium directly inhibits IMPase activity by noncompetitively displacing Mg^{2+} from the enzyme's catalytic sites, reducing inositol triphosphate formation [55]. This, in turn, modulates several intracellular pathways relevant to neuropsychiatric disorders, particularly by stimulating autophagy [56]. The autophagic effect of lithium may facilitate the clearance of aggregation-prone proteins, such as Aβ, phospho-tau (p-tau), and damaged mitochondria, potentially influencing the pathogenesis of AD [57,58].

GSK-3β is a constitutively active enzyme involved in the organization and remodeling of the cytoskeleton [59,60]. Lithium inhibits GSK-3β activity through two primary mechanisms: (a) directly, by competing with Mg^{2+} ions for binding to the enzyme's catalytic site, and (b) indirectly, by inducing phosphorylation of the serine-9 residue of GSK-3β, leading to conformational changes that inactivate the enzyme [53,54,58]. These indirect mechanisms activate intracellular kinases, such as Akt, while inhibiting the intracellular protein phosphatase 2A (PP2A) [60]. GSK-3β inhibition also interferes with the Wnt/β-catenin signaling pathway, leading to the intracellular accumulation of β-catenin, which facilitates its entry into the nucleus to regulate the transcription of target genes and modulates several downstream pathological processes [60–62]. Ultimately, this reduces Aβ deposition, enhances neurogenesis, and improves mitochondrial bioenergetics [53,63–65]. Additionally, lithium can inhibit the mRNA transcription of GSK-3β, further reducing the enzyme's availability [58,66].

Neuroinflammation plays a crucial role in neurodegenerative diseases. The overactivation of microglia and astrocytes, along with the inflammatory molecules they produce, can disrupt the neuronal microenvironment and lead to cognitive impairment [67]. When stimulated by immune responses or tissue damage, reactive pro-inflammatory microglia, activated by interferon-γ and tumor necrosis factor-α (TNF-α), release pro-inflammatory cytokines such as interleukin-1β (IL-1β), IL-6, IL-18, and TNF-α, along with nitric oxide (NO) and reactive oxygen species (ROS). These substances are associated with increased neurotoxicity [67,68]. Lithium treatment has the potential to reduce the production of pro-inflammatory factors, including IL-1β and TNF-α, as well as other neuroinflammatory biomarkers in animal models of AD [41,42,49,63,69].

9. Exploring Lithium's Neuroprotective Potential

Evidence of lithium's neuroprotective effects comes from experimental models and neuroimaging studies. Chronic lithium use in patients with BD is associated with an increase in the volume of cerebral gray matter and improved tissue viability [70–72]. Furthermore, these patients show a decreased prevalence of dementia compared to patients treated with other mood stabilizers. This benefit appears to be independent of therapeutic response parameters, suggesting that it may be due to lithium's effects on brain systems responsible for maintaining homeostasis, including second messenger pathways, cellular communication, and cell viability [73].

Lithium-mediated inhibition of GSK-3β is causally linked to its effects on key pathogenic pathways of AD, particularly the Aβ cascade and hyperphosphorylated tau (p-tau). The accumulation of Aβ and p-tau is responsible for the formation of amyloid plaques and neurofibrillary tangles (NFTs), which are the key pathological hallmarks of AD. In the Aβ cascade, the APP can be cleaved by α-secretases, which generate non-toxic APP fragments, or by β- and γ-secretases, which produce toxic peptides through the amyloidogenic pathway. As previously mentioned, elevated levels of nuclear β-catenin, caused by GSK-3β inhibition, result in its interaction with transcription factor 4 (TCF4). This interaction inhibits the transcription of the main enzyme related to APP cleavage at the β1 site (BACE1), ultimately reducing Aβ production [49,60]. Additionally, lithium attenuates the activity of γ-secretase and enhances the clearance of Aβ by improving the function of the blood–brain barrier (BBB) microvessels, promoting Aβ efflux through cerebrospinal fluid (CSF) bulk flow [74].

The deposition of hyperphosphorylated tau protein (p-tau) leads to the formation of intraneuronal paired helical filaments and NFTs. This process disrupts the microtubule structure, impairs normal axonal transport of nutrients and signaling, and ultimately results in cell death. Lithium may help prevent the progression of tau neuropathology in AD by inhibiting GSK-3β and modifying the expression or activity of other tau kinases, such as protein kinase A (PKA), Akt (PKB), and calcium/calmodulin-dependent kinase II (CaMKII) [60,66,75]. Furthermore, lithium increases the expression of B-cell leukemia 2 family protein (Bcl-2), a cytoprotective protein that promotes axon regeneration by inhibiting apoptosis and supporting the remodeling of the neuronal cytoskeleton [76]. In experimental models, lithium has been shown to stimulate neurogenesis in the dentate gyrus of the hippocampus. Among lithium's neuroprotective effects, the stimulation of the synthesis and release of neurotrophins, such as brain-derived neurotrophic factor (BDNF) and vascular endothelial growth factor (VEGF), is also noteworthy [42,49,58]. Figure 1 illustrates the mechanism through which lithium exerts neuroprotection against the neuropathogenic processes of AD.

In summary, the neuroprotective actions of lithium against neuropathogenic mechanisms of AD and other neurodegenerative disorders are partly due to its intrinsic effects on oxidative stress, autophagy, apoptosis, mitochondrial function, and neuroinflammation. Evidence supporting this hypothesis has emerged from the outcomes of translational research [73]. Controlled studies demonstrated that prolonged use of lithium carbonate in subtherapeutic doses could slow cognitive and functional decline while reducing the transition to dementia in a population. Notably, this clinical effect was accompanied by changes in the profile of AD CSF biomarkers [77–79].

Lithium is being explored as a potential disease-modifying agent/treatment not only for BD and AD but also for other neurodegenerative disorders such as Parkinson's disease [41], Huntington's disease [80], amyotrophic lateral sclerosis [81], and stroke [42]. The regulatory effects of lithium can vary based on factors such as the target tissue, duration of exposure, and tissue concentration at the site of action. Consequently, some biological effects may occur at lower tissue concentrations than those typically encountered in the treatment of mood disorders. Ecological studies conducted globally have shown an inverse relationship between suicide rates and lithium concentrations in groundwater, drinking water, and clinical settings [82,83]. Similar associations have been observed between environmental lithium and psychiatric hospitalization rates, incidents of violence [84,85] and the prevalence of dementia [36]. These findings suggest that the intake of lithium in low doses may result in long-term beneficial outcomes [43,77,78,83,86–89].

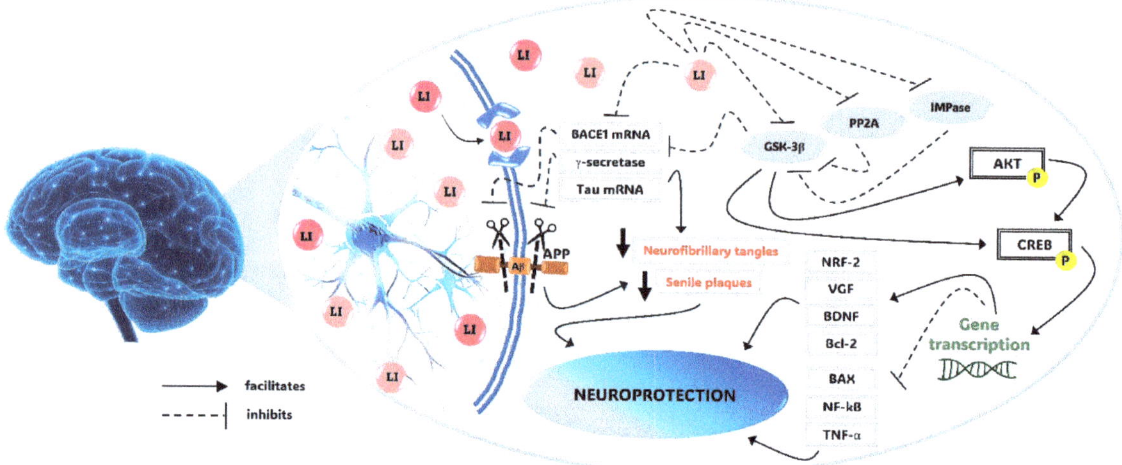

Figure 1. Molecular mechanisms of lithium in Alzheimer's Disease. Alzheimer's disease (AD) is characterized by the accumulation of senile plaques and neurofibrillary tangles (NFTs) in the brain, disrupting normal neuronal function. Lithium modulates the production and clearance of amyloid-beta (Aβ) by inhibiting β-site APP cleaving enzyme 1 (BACE1) mRNA expression and decreasing γ-secretase activity, thereby reducing amyloid precursor protein (APP) cleavage and Aβ production. Additionally, lithium reduces NFT formation by lowering tau mRNA expression and inhibiting glycogen synthase kinase-3 beta (GSK-3β), which in turn decreases tau phosphorylation levels and results in neuroprotection. At the same time, through the inhibition of GSK-3β, lithium induces the activation of protein kinase B (Akt) and cyclic adenosine monophosphate (cAMP) response element-binding protein (CREB) signaling pathways, thereby upregulating the mRNA expression and protein levels of nuclear factor erythroid 2-related factor 2 (NRF-2), neurosecretory protein VGF, brain-derived neurotrophic factor (BDNF), B-cell lymphoma 2 (Bcl-2), and other neuroprotective molecules. Conversely, by this same pathway, lithium downregulates the mRNA expression and protein levels of Bcl-2–associated X protein (BAX), nuclear factor kappa-light-chain-enhancer of activated B cells (NF-κB), tumor necrosis factor-alpha (TNF-α), and other molecules involved in neuroinflammation and apoptosis. Furthermore, by inhibiting protein phosphatase 2 (PP2A) and inositol monophosphatase (IMPase), lithium reinforces GSK-3β inactivation, promoting neuroprotective responses.

10. Conclusions

The evolution of lithium therapy highlights the progression of psychiatric medicine, from initial skepticism to its establishment as an essential therapeutic tool, underscoring the importance of continued research and clinical advancements in mental health treatment. Research from both preclinical and clinical models supports the idea that lithium has neurotrophic and neuroprotective effects. These beneficial effects may be due to its effects on various cellular metabolic processes crucial for neuronal survival, neuroplasticity, transcriptional regulation, energy metabolism, and protection against neurotoxic damage. Some of these mechanisms are directly associated with the central pathogenic processes of specific diseases, while others may reflect general responses that enhance intrinsic neuronal resilience and foster neuroprotective effects.

It is important to note that the clinical use of lithium should be limited to conditions where its efficacy has been well established by scientific evidence, such as mood disorders. For other health conditions, further research is needed to support new therapeutic indications. Lithium treatments, including off-label use, should consider factors such as dosage, the specific health condition, age, and potential drug interactions.

In light of this, it is evident that future research should aim to further explore the biochemical mechanisms of lithium. Investigating the molecular pathways it influences will give us insight into its clinical efficacy and interaction with cellular processes. This enhanced knowledge will improve lithium's current use in mood disorders and possibly reveal new therapeutic targets. A more profound understanding of lithium's mechanisms will help develop more precise treatment strategies and novel pharmacological approaches.

Author Contributions: M.R.: Conceptualization, Writing—Original Draft; M.P.S.: Conceptualization, Writing—Original Draft, Writing—Review and Editing, Production of Graphical Abstract; V.d.J.R.D.P.: Writing—Original Draft, Writing—Review and Editing; L.L.T.: Writing—Review and Editing; O.V.F.: Supervision, Conceptualization, Writing—Review and Editing, Funding Acquisition. All authors have read and agreed to the published version of the manuscript.

Funding: The study was funded by the São Paulo Research Foundation (FAPESP; grant number 2021/06378-1), Programa Nacional de Apoio à Atenção da Saúde da Pessoa com Deficiência (PRONAS; grant number 25000.002058/2020-71), and Instituto Nacional de Biomarcadores em Neuropsiquiatria, Conselho Nacional de Desenvolvimento Científico e Tecnológico (INBioN; grant numbers 465412/2014-9 and 2014/50873-3).

Data Availability Statement: No new data were created or analyzed in this study. Data sharing is not applicable to this article.

Acknowledgments: We are grateful to the funding agencies Fundação Coordenação de Aperfeiçoamento de Pessoal de Nível Superior (CAPES), Conselho Nacional de Desenvolvimento Científico e Tecnológico (CNPq), and FAPESP.

Conflicts of Interest: The authors declare no conflict of interest.

References

1. Bernard, A. *Lithium. Handbook on the Toxicology of Metals*; Elsevier: Amsterdam, The Netherlands, 2015; Volume 2, pp. 969–974, ISBN 978-0-444-59453-2.
2. Lazarus, J.H.; Collard, K.J. Lithium—History and Pharmacology. In *Endocrine and Metabolic Effects of Lithium*; Lazarus, J.H., Collard, K.J., Eds.; Springer US: Boston, MA, USA, 1986; pp. 1–15, ISBN 978-1-4899-2016-4.
3. Cousins, D.A.; Squarcina, L.; Boumezbeur, F.; Young, A.H.; Bellivier, F. Lithium: Past, Present, and Future. *Lancet Psychiatry* **2020**, *7*, 222–224. [CrossRef] [PubMed]
4. Shorter, E. The History of Lithium Therapy. *Bipolar Disord.* **2009**, *11*, 4–9. [CrossRef] [PubMed]
5. Johnson, F.N. *The History of Lithium Therapy*, 1st ed.; Palgrave Macmillan: London, UK, 1984; ISBN 978-1-349-07289-7.
6. Airainer, M.; Seifert, R. Lithium, the Gold Standard Drug for Bipolar Disorder: Analysis of Current Clinical Studies. *Naunyn-Schmiedeberg's Arch. Pharmacol.* **2024**, *397*, 9723–9743. [CrossRef] [PubMed]
7. Chagas, C.; Corrêa, T. As Contribuições Científicas de José Bonifácio e a Descoberta Do Lítio: Um Caminhar Pela História Da Ciência. *Rev. Educ. Ciênc. Mat.* **2017**, *7*, 201–212.
8. Fountoulakis, K.N. Historical Perspective. In *Bipolar Disorder: An Evidence-Based Guide to Manic Depression*; Fountoulakis, K.N., Ed.; Springer: Berlin/Heidelberg, Germany, 2015; pp. 1–25, ISBN 978-3-642-37216-2.
9. Garrod, A.B. *The Nature and Treatment of Gout and Rheumatic Gout*; Walton and Maberly: London, UK, 1859.
10. Haig, A. The Uric Acid Diathesis. *Br. Med. J.* **1896**, *2*, 1746–1747. [CrossRef]
11. Hammond, W.A. A Treatise on Diseases of the Nervous System. Appleton and Co.: New York, NY, USA, 1871; pp. 32–364, ISBN 0282051856.
12. Schioldann, J. 'On periodical depressions and their pathogenesis' by Carl Lange (1886). *Hist. Psychiatry* **2011**, *22*, 108–130. [CrossRef]
13. Baldessarini, R.J.; Stephens, J.H. Lithium Carbonate for Affective Disorders. I. Clinical Pharmacology and Toxicology. *Arch. Gen. Psychiatry* **1970**, *22*, 72–77. [CrossRef]
14. Culbreth, D.M. *A Manual of Materia Medica and Pharmacology: Comprising All Organic and Inorganic Drugs Which Are or Have Been Official in the United States Pharmacopoeia, Together with Important Allied Species and Useful Synthetics, Especially Designed for Students of Pharmacy and Medicine, as Well as for Druggists, Pharmacists, and Physicians*; Lea & Febiger: Philadelphia, PA, USA, 1906.
15. Cade, J.F.; Malhi, G.S. Cade's Lithium. *Acta Neuropsychiatr.* **2007**, *19*, 125–126. [CrossRef]
16. Cade, J.F.J. Lithium Salts in the Treatment of Psychotic Excitement. *Med. J. Aust.* **1949**, *2*, 349–352. [CrossRef]

17. Burrows, G.D.; Tiller, J.W. Cade's Observation of the Antimanic Effect of Lithium and Early Australian Research. *Aust. N. Z. J. Psychiatry* **1999**, *33*, S27–S31. [CrossRef]
18. Johnson, F.N. The Work of John Cade. In *The History of Lithium Therapy*; Palgrave Macmillan UK: London, UK, 1984; pp. 34–45, ISBN 978-1-349-07291-0.
19. Noack, C.H.; Trautner, E.M. The Lithium Treatment of Maniacal Psychosis. *Med. J. Aust.* **1951**, *2*, 219–222. [CrossRef]
20. Kline, N.S. *Depression: Its Diagnosis and Treatment—Lithium: The History of Its Use in Psychiatry*; Karger Medical and Scientific Publishers: Basel, Switzerland, 1969; ISBN 978-3-318-03958-0.
21. Trautner, E.M.; Morris, R.; Noack, C.H.; Gershon, S. The excretion and retention of ingested lithium and its effect on the ionic balance of man. *Med. J. Aust.* **1955**, *2*, 280–291. [CrossRef]
22. Schou, M.; Juel-Nielsen, N.; Stromgren, E.; Voldby, H. The Treatment of Manic Psychoses by the Administration of Lithium Salts. *J. Neurol. Neurosurg. Psychiatry* **1954**, *17*, 250–260. [CrossRef] [PubMed]
23. Baastrup, P.C.; Schou, M. Lithium as a Prophylactic Agents. Its Effect against Recurrent Depressions and Manic-Depressive Psychosis. *Arch. Gen. Psychiatry* **1967**, *16*, 162–172. [CrossRef] [PubMed]
24. Baastrup, P.C.; Poulsen, J.C.; Schou, M.; Thomsen, K.; Amdisen, A. Prophylactic lithium: Double blind discontinuation in manic-depressive and recurrent-depressive disorders. *Lancet* **1970**, *296*, 326–330. [CrossRef]
25. Schou, M.; Amdisen, A.; Baastrup, P.C. The Practical Management of Lithium Treatment. *Br. J. Psychiatry* **1975**, 76–84.
26. Schou, M. Current Status of Lithium Therapy in Affective Disorders. *Arq. Neuro-Psiquiatr.* **1976**, *34*, 68–80. [CrossRef]
27. Gershon, S.; Yuwiler, A. Lithium Ion: A Specific Psychopharmacological Approach to the Treatment of Mania. *J. Neuropsychiatr.* **1960**, *1*, 229–241.
28. Gershon, S. *Lithium: Its Role in Psychiatric Research and Treatment*, 1st ed.; Plenum Press: New York, NY, USA, 1973; ISBN 978-0-306-30720-1.
29. Johnson, G.; Gershon, S. Early North American Research on Lithium. *Aust. N. Z. J. Psychiatry* **1999**, *33*, S48–S53. [CrossRef]
30. The Current Status of Lithium Therapy: Report of the APA Task Force. *Am. J. Psychiatry* **1975**, *132*, 977–1001. [CrossRef]
31. Squire, L.R.; Judd, L.L.; Janowsky, D.S.; Huey, L.Y. Effects of Lithium Carbonate on Memory and Other Cognitive Functions. *Am. J. Psychiatry* **1980**, *137*, 1042–1046. [CrossRef]
32. Reus, V.I.; Targum, S.D.; Weingartner, H.; Post, R.M. Effect of Lithium Carbonate on Memory Processes of Bipolar Affectively Ill Patients. *Psychopharmacology* **1979**, *63*, 39–42. [CrossRef]
33. Kusumo, K.S.; Vaughan, M. Effects of Lithium Salts on Memory. *Br. J. Psychiatry* **1977**, *131*, 453–457. [CrossRef] [PubMed]
34. Rybakowski, J.K. Effect of Lithium on Neurocognitive Functioning. *Curr. Alzheimer Res.* **2016**, *13*, 887–893. [CrossRef]
35. Manchia, M.; Paribello, P.; Pinna, M.; Steardo, L.; Carpiniello, B.; Pinna, F.; Pisanu, C.; Squassina, A.; Hajek, T. Lithium and Its Effects: Does Dose Matter? *Int. J. Bipolar Disord.* **2024**, *12*, 23. [CrossRef] [PubMed]
36. Gitlin, M.; Bauer, M. Lithium: Current State of the Art and Future Directions. *Int. J. Bipolar Disord.* **2024**, *12*, 40. [CrossRef] [PubMed]
37. Williams, K.H.; Goldstein, G. Cognitive and Affective Responses to Lithium in Patients with Organic Brain Syndrome. *Am. J. Psychiatry* **1979**, *136*, 800–803. [CrossRef]
38. Calil, H.M.; Zwicker, A.P.; Klepacz, S. The Effects of Lithium Carbonate on Healthy Volunteers: Mood Stabilization? *Biol. Psychiatry* **1990**, *27*, 711–722. [CrossRef]
39. Clarke, N.A.; Procter, A.W.; Webster, M.T.; Francis, P.T.; Hodgkiss, A.D.; Bowen, D.M. Effect of Li+ Therapy and Site of Origin on Cerebrospinal Fluid Beta-Amyloid Precursor Protein Derivatives. *Biochem. Soc. Trans.* **1993**, *21 Pt 3*, 242S. [CrossRef]
40. Grof, P. Sixty Years of Lithium Responders. *Neuropsychobiology* **2010**, *62*, 8–16. [CrossRef] [PubMed]
41. Singulani, M.P.; Ferreira, A.F.F.; Figueroa, P.S.; Cuyul-Vásquez, I.; Talib, L.L.; Britto, L.R.; Forlenza, O.V. Lithium and Disease Modification: A Systematic Review and Meta-Analysis in Alzheimer's and Parkinson's Disease. *Ageing Res. Rev.* **2024**, *95*, 102231. [CrossRef] [PubMed]
42. Almeida, O.P.; Singulani, M.P.; Ford, A.H.; Hackett, M.L.; Etherton-Beer, C.; Flicker, L.; Hankey, G.J.; De Paula, V.J.R.; Penteado, C.T.; Forlenza, O.V. Lithium and Stroke Recovery: A Systematic Review and Meta-Analysis of Stroke Models in Rodents and Human Data. *Stroke* **2022**, *53*, 2935–2944. [CrossRef] [PubMed]
43. Nunes, M.A.; Viel, T.A.; Buck, H.S. Microdose Lithium Treatment Stabilized Cognitive Impairment in Patients with Alzheimer's Disease. *Curr. Alzheimer Res.* **2013**, *10*, 104–107. [CrossRef] [PubMed]
44. Iordache, A.M.; Voica, C.; Roba, C.; Nechita, C. Lithium Content and Its Nutritional Beneficence, Dietary Intake, and Impact on Human Health in Edibles from the Romanian Market. *Foods* **2024**, *13*, 592. [CrossRef]
45. El-Mallakh, R.S. Ion Homeostasis and the Mechanism of Action of Lithium. *Clin. Neurosci. Res.* **2004**, *4*, 227–231. [CrossRef]
46. Kakhki, S.; Ahmadi-Soleimani, S.M. Experimental Data on Lithium Salts: From Neuroprotection to Multi-Organ Complications. *Life Sci.* **2022**, *306*, 120811. [CrossRef]
47. Schrauzer, G.N. Lithium: Occurrence, Dietary Intakes, Nutritional Essentiality. *J. Am. Coll. Nutr.* **2002**, *21*, 14–21. [CrossRef]

48. Ramos, P.; Santos, A.; Pinto, E.; Pinto, N.R.; Mendes, R.; Magalhães, T.; Almeida, A. Alkali Metals Levels in the Human Brain Tissue: Anatomical Region Differences and Age-Related Changes. *J. Trace Elem. Med. Biol.* **2016**, *38*, 174–182. [CrossRef]
49. Shen, Y.; Zhao, M.; Zhao, P.; Meng, L.; Zhang, Y.; Zhang, G.; Taishi, Y.; Sun, L. Molecular Mechanisms and Therapeutic Potential of Lithium in Alzheimer's Disease: Repurposing an Old Class of Drugs. *Front. Pharmacol.* **2024**, *15*, 1408462. [CrossRef]
50. Wesseloo, R.; Wierdsma, A.I.; van Kamp, I.L.; Munk-Olsen, T.; Hoogendijk, W.J.G.; Kushner, S.A.; Bergink, V. Lithium Dosing Strategies during Pregnancy and the Postpartum Period. *Br. J. Psychiatry* **2017**, *211*, 31–36. [CrossRef]
51. Mohandas, E.; Rajmohan, V. Lithium Use in Special Populations. *Indian J. Psychiatry* **2007**, *49*, 211–218. [CrossRef] [PubMed]
52. Tondo, L.; Alda, M.; Bauer, M.; Bergink, V.; Grof, P.; Hajek, T.; Lewitka, U.; Licht, R.W.; Manchia, M.; Müller-Oerlinghausen, B.; et al. Clinical Use of Lithium Salts: Guide for Users and Prescribers. *Int. J. Bipolar Disord.* **2019**, *7*, 16. [CrossRef]
53. Puglisi-Allegra, S.; Ruggieri, S.; Fornai, F. Translational Evidence for Lithium-Induced Brain Plasticity and Neuroprotection in the Treatment of Neuropsychiatric Disorders. *Transl. Psychiatry* **2021**, *11*, 1–10. [CrossRef] [PubMed]
54. Gogoleva, I.V.; Gromova, O.A.; Torshin, I.Y.; Grishina, T.R.; Pronin, A.V. The Neurobiological Role of Lithium Salts. *Neurosci. Behav. Physi* **2023**, *53*, 939–945. [CrossRef]
55. Patel, S.; Yenush, L.; Rodríguez, P.L.; Serrano, R.; Blundell, T.L. Crystal Structure of an Enzyme Displaying Both Inositol-Polyphosphate-1-Phosphatase and 3′-Phosphoadenosine-5′-Phosphate Phosphatase Activities: A Novel Target of Lithium Therapy. *J. Mol. Biol.* **2002**, *315*, 677–685. [CrossRef]
56. Sarkar, S.; Floto, R.A.; Berger, Z.; Imarisio, S.; Cordenier, A.; Pasco, M.; Cook, L.J.; Rubinsztein, D.C. Lithium Induces Autophagy by Inhibiting Inositol Monophosphatase. *J. Cell Biol.* **2005**, *170*, 1101–1111. [CrossRef]
57. Singulani, M.P.; De Paula, V.J.R.; Forlenza, O.V. Mini Review: Mitochondrial Dysfunction in Alzheimer's Disease: Therapeutic Implications of Lithium. *Neurosci. Lett.* **2021**, *760*, 136078. [CrossRef] [PubMed]
58. Forlenza, O.V.; De-Paula, V.J.R.; Diniz, B.S.O. Neuroprotective Effects of Lithium: Implications for the Treatment of Alzheimer's Disease and Related Neurodegenerative Disorders. *ACS Chem. Neurosci.* **2014**, *5*, 443–450. [CrossRef]
59. Hajka, D.; Budziak, B.; Pietras, Ł.; Duda, P.; McCubrey, J.A.; Gizak, A. GSK3 as a Regulator of Cytoskeleton Architecture: Consequences for Health and Disease. *Cells* **2021**, *10*, 2092. [CrossRef]
60. Chiu, C.-T.; Chuang, D.-M. Molecular Actions and Therapeutic Potential of Lithium in Preclinical and Clinical Studies of CNS Disorders. *Pharmacol. Ther.* **2010**, *128*, 281–304. [CrossRef]
61. Nusse, R. Wnt Signaling in Disease and in Development. *Cell Res.* **2005**, *15*, 28–32. [CrossRef] [PubMed]
62. De-Paula, V.J.; dos Santos, C.C.C.; Luque, M.C.A.; Ali, T.M.; Kalil, J.E.; Forlenza, O.V.; Cunha-Neto, E. Acute and Chronic Lithium Treatment Increases Wnt/β-Catenin Transcripts in Cortical and Hippocampal Tissue at Therapeutic Concentrations in Mice. *Metab. Brain Dis.* **2021**, *36*, 193–197. [CrossRef] [PubMed]
63. Wilson, E.N.; Do Carmo, S.; Iulita, M.F.; Hall, H.; Ducatenzeiler, A.; Marks, A.R.; Allard, S.; Jia, D.T.; Windheim, J.; Cuello, A.C. BACE1 Inhibition by Microdose Lithium Formulation NP03 Rescues Memory Loss and Early Stage Amyloid Neuropathology. *Transl. Psychiatry* **2017**, *7*, e1190. [CrossRef] [PubMed]
64. Fiorentini, A.; Rosi, M.C.; Grossi, C.; Luccarini, I.; Casamenti, F. Lithium Improves Hippocampal Neurogenesis, Neuropathology and Cognitive Functions in APP Mutant Mice. *PLoS ONE* **2010**, *5*, e14382. [CrossRef]
65. Puglisi-Allegra, S.; Lazzeri, G.; Busceti, C.L.; Giorgi, F.S.; Biagioni, F.; Fornai, F. Lithium Engages Autophagy for Neuroprotection and Neuroplasticity: Translational Evidence for Therapy. *Neurosci. Biobehav. Rev.* **2023**, *148*, 105148. [CrossRef]
66. Mendes, C.T.; Mury, F.B.; de Sá Moreira, E.; Alberto, F.L.; Forlenza, O.V.; Dias-Neto, E.; Gattaz, W.F. Lithium Reduces Gsk3b mRNA Levels: Implications for Alzheimer Disease. *Eur. Arch. Psychiatry Clin. Neurosci.* **2009**, *259*, 16–22. [CrossRef]
67. Stanca, S.; Rossetti, M.; Bongioanni, P. Astrocytes as Neuroimmunocytes in Alzheimer's Disease: A Biochemical Tool in the Neuron–Glia Crosstalk along the Pathogenetic Pathways. *Int. J. Mol. Sci.* **2023**, *24*, 13880. [CrossRef]
68. Damri, O.; Agam, G. Lithium, Inflammation and Neuroinflammation with Emphasis on Bipolar Disorder-A Narrative Review. *Int. J. Mol. Sci.* **2024**, *25*, 13277. [CrossRef]
69. Wilson, E.N.; Do Carmo, S.; Welikovitch, L.A.; Hall, H.; Aguilar, L.F.; Foret, M.K.; Iulita, M.F.; Jia, D.T.; Marks, A.R.; Allard, S.; et al. NP03, a Microdose Lithium Formulation, Blunts Early Amyloid Post-Plaque Neuropathology in McGill-R-Thy1-APP Alzheimer-Like Transgenic Rats. *J. Alzheimer's Dis.* **2020**, *73*, 723–739. [CrossRef]
70. Bora, E.; Fornito, A.; Yücel, M.; Pantelis, C. Voxelwise Meta-Analysis of Gray Matter Abnormalities in Bipolar Disorder. *Biol. Psychiatry* **2010**, *67*, 1097–1105. [CrossRef]
71. Lyoo, I.K.; Dager, S.R.; Kim, J.E.; Yoon, S.J.; Friedman, S.D.; Dunner, D.L.; Renshaw, P.F. Lithium-Induced Gray Matter Volume Increase As a Neural Correlate of Treatment Response in Bipolar Disorder: A Longitudinal Brain Imaging Study. *Neuropsychopharmacology* **2010**, *35*, 1743–1750. [CrossRef] [PubMed]
72. Ferensztajn-Rochowiak, E.; Chłopocka-Woźniak, M.; Rybakowski, J.K. Ultra-Long-Term Lithium Therapy: All-Important Matters and a Case of Successful 50-Year Lithium Treatment. *Braz. J. Psychiatry* **2020**, *43*, 407–413. [CrossRef]

73. Kessing, L.V.; Gerds, T.A.; Knudsen, N.N.; Jørgensen, L.F.; Kristiansen, S.M.; Voutchkova, D.; Ernstsen, V.; Schullehner, J.; Hansen, B.; Andersen, P.K.; et al. Lithium in Drinking Water and the Incidence of Bipolar Disorder: A Nation-Wide Population-Based Study. *Bipolar Disord.* **2017**, *19*, 563–567. [CrossRef]
74. Pan, Y.; Short, J.L.; Newman, S.A.; Choy, K.H.C.; Tiwari, D.; Yap, C.; Senyschyn, D.; Banks, W.A.; Nicolazzo, J.A. Cognitive Benefits of Lithium Chloride in APP/PS1 Mice Are Associated with Enhanced Brain Clearance of β-Amyloid. *Brain Behav. Immun.* **2018**, *70*, 36–47. [CrossRef] [PubMed]
75. De-Paula, V.J.; Forlenza, O.V. Lithium Modulates Multiple Tau Kinases with Distinct Effects in Cortical and Hippocampal Neurons According to Concentration Ranges. *Naunyn Schmiedebergs Arch. Pharmacol.* **2022**, *395*, 105–113. [CrossRef]
76. Manji, H.K.; Moore, G.J.; Chen, G. Lithium Up-Regulates the Cytoprotective Protein Bcl-2 in the CNS in Vivo: A Role for Neurotrophic and Neuroprotective Effects in Manic Depressive Illness. *J. Clin. Psychiatry* **2000**, *61* (Suppl. S9), 82–96.
77. Forlenza, O.V.; Diniz, B.S.; Radanovic, M.; Santos, F.S.; Talib, L.L.; Gattaz, W.F. Disease-Modifying Properties of Long-Term Lithium Treatment for Amnestic Mild Cognitive Impairment: Randomised Controlled Trial. *Br. J. Psychiatry* **2011**, *198*, 351–356. [CrossRef]
78. Forlenza, O.V.; Radanovic, M.; Talib, L.L.; Gattaz, W.F. Clinical and Biological Effects of Long-Term Lithium Treatment in Older Adults with Amnestic Mild Cognitive Impairment: Randomised Clinical Trial. *Br. J. Psychiatry* **2019**, *215*, 668–674. [CrossRef]
79. Hampel, H.; Ewers, M.; Bürger, K.; Annas, P.; Mörtberg, A.; Bogstedt, A.; Frölich, L.; Schröder, J.; Schönknecht, P.; Riepe, M.W.; et al. Lithium Trial in Alzheimer's Disease: A Randomized, Single-Blind, Placebo-Controlled, Multicenter 10-Week Study. *J. Clin. Psychiatry* **2009**, *70*, 922–931.
80. Danivas, V.; Moily, N.S.; Thimmaiah, R.; Muralidharan, K.; Purushotham, M.; Muthane, U.; Jain, S. Off Label Use of Lithium in the Treatment of Huntington's Disease: A Case Series. *Indian J. Psychiatry* **2013**, *55*, 81–83. [CrossRef]
81. UKMND-LiCALS Study Group; Morrison, K.E.; Dhariwal, S.; Hornabrook, R.; Savage, L.; Burn, D.J.; Khoo, T.K.; Kelly, J.; Murphy, C.L.; Al-Chalabi, A.; et al. Lithium in Patients with Amyotrophic Lateral Sclerosis (LiCALS): A Phase 3 Multicentre, Randomised, Double-Blind, Placebo-Controlled Trial. *Lancet Neurol.* **2013**, *12*, 339–345. [CrossRef] [PubMed]
82. Liaugaudaite, V.; Mickuviene, N.; Raskauskiene, N.; Naginiene, R.; Sher, L. Lithium Levels in the Public Drinking Water Supply and Risk of Suicide: A Pilot Study. *J. Trace Elem. Med. Biol.* **2017**, *43*, 197–201. [CrossRef] [PubMed]
83. Fraiha-Pegado, J.; de Paula, V.J.R.; Alotaibi, T.; Forlenza, O.; Hajek, T. Trace Lithium Levels in Drinking Water and Risk of Dementia: A Systematic Review. *Int. J. Bipolar Disord.* **2024**, *12*, 32. [CrossRef]
84. Müller-Oerlinghausen, B.; Lewitzka, U. Lithium Reduces Pathological Aggression and Suicidality: A Mini-Review. *Neuropsychobiology* **2010**, *62*, 43–49. [CrossRef]
85. Lewitzka, U.; Severus, E.; Bauer, R.; Ritter, P.; Müller-Oerlinghausen, B.; Bauer, M. The Suicide Prevention Effect of Lithium: More than 20 Years of Evidence-a Narrative Review. *Int. J. Bipolar Disord.* **2015**, *3*, 32. [CrossRef] [PubMed]
86. Devanand, D.P.; Crocco, E.; Forester, B.P.; Husain, M.M.; Lee, S.; Vahia, I.V.; Andrews, H.; Simon-Pearson, L.; Imran, N.; Luca, L.; et al. Low Dose Lithium Treatment of Behavioral Complications in Alzheimer's Disease: Lit-AD Randomized Clinical Trial. *Am. J. Geriatr. Psychiatry* **2022**, *30*, 32–42. [CrossRef]
87. Devanand, D.P.; Pelton, G.H.; D'Antonio, K.; Strickler, J.G.; Kreisl, W.C.; Noble, J.; Marder, K.; Skomorowsky, A.; Huey, E.D. Low-Dose Lithium Treatment for Agitation and Psychosis in Alzheimer Disease and Frontotemporal Dementia: A Case Series. *Alzheimer Dis. Assoc. Disord.* **2017**, *31*, 73–75. [CrossRef]
88. Devanand, D.P.; Strickler, J.G.; Huey, E.D.; Crocco, E.; Forester, B.P.; Husain, M.M.; Vahia, I.V.; Andrews, H.; Wall, M.M.; Pelton, G.H. Lithium Treatment for Agitation in Alzheimer's Disease (Lit-AD): Clinical Rationale and Study Design. *Contemp. Clin. Trials* **2018**, *71*, 33–39. [CrossRef]
89. Damiano, R.F.; Loureiro, J.C.; Pais, M.V.; Pereira, R.F.; Corradi, M.D.M.; Santi, T.D.; Bezerra, G.A.M.; Radanovic, M.; Talib, L.L.; Forlenza, O.V. Revisiting Global Cognitive and Functional State 13 Years after a Clinical Trial of Lithium for Mild Cognitive Impairment. *Braz. J. Psychiatry* **2023**, *45*, 46–49. [CrossRef]

pharmaceuticals

MDPI

Review

Trace Lithium for Suicide Prevention and Dementia Prevention: A Qualitative Review

Takeshi Terao *, Hirofumi Hirakawa, Masaaki Muronaga, Toshihiko Izumi and Kentaro Kohno

Department of Neuropsychiatry, Oita University Faculty of Medicine, Daigaoka 1-1, Hasama-machi, Yufu 879-5593, Oita, Japan; hira-hiro@oita-u.ac.jp (H.H.); m-muronaga@oita-u.ac.jp (M.M.); tshhkizm@oita-u.ac.jp (T.I.); kentarokohno@oita-u.ac.jp (K.K.)

* Correspondence: terao@oita-u.ac.jp

Abstract: Background: Anti-manic effects of lithium and the nature of trace element in lithium were first observed in 1949. In this review, we explore the potential effects of trace lithium in the prevention of suicide and dementia. **Methods**: This is a qualitative and non-systematic review. **Results**: While most studies to date have been cross-sectional, limiting the establishment of causal relationships, the potential benefits of trace lithium for suicide prevention and dementia prevention are notable, especially in the absence of radical treatments for suicide and dementia. Furthermore, trace lithium appears to lack many of the adverse effects associated with so-called therapeutic lithium levels. **Conclusions**: The present findings suggest that trace lithium may be associated with lower suicide rates and reduced dementia rates. Probably, trace lithium may inhibit testosterone and thereby mitigate aggression and impulsivity and decrease suicide. Also, trace lithium may inhibit GSK-3β and thereby lower amyloid β and tau hyperphosphorylation and inhibit pro-inflammatory cytokines such as IL 6 and IL 8 and thereby mitigate inflammation, whereas trace lithium may promote BDNF and neurogenesis in the general population.

Keywords: lithium; suicide prevention; dementia; trace elements

1. Introduction

For the first time, Cade [1] identified the therapeutic effects of lithium effects for bipolar disorder, noting, "There is no doubt that in mania patients improvement had closely paralleled (lithium) treatment" and "The quietening effect on restless non-manic psychotics is additional strong evidence of the efficacy of lithium salts, especially as such restlessness returned on cessation of (lithium) treatment". Subsequent studies have repeatedly confirmed lithium's mood-stabilizing properties in bipolar disorder [2–4], as well as its efficacy in augmentation therapy for treatment-resistant major depression [5–7]. Notably, several investigations have demonstrated lithium's anti-suicidal effects [8,9] and potential anti-dementia properties [10–12]. Cade [1] also speculated, "Lithium may well be an essential trace element. It is widely distributed, has been detected in sea-water and in many spring and river waters, in the ash of many plants, and in animal ash".

As for the clinical use of lithium for bipolar disorder, recently Shuy et al. [13] reported that lithium salts were prescribed in 27.3% of 2139 patients suffering from bipolar disorder (52.3% women, mean age 42.4 years), varying among regions from 3.20% to 59.5%. Associated with lithium treatment were male sex, the presence of euthymia or mild depression, and a history of seasonal mood change. Other mood stabilizers were usually given with lithium, often at relatively high doses. Lithium use was associated with a newly emerging and dose-dependent risk of tremors as well as a risk of hypothyroidism. Shuy et al. [13] found no significant differences in rates of clinical remission or of suicidal behavior whether treatment included lithium or not. As for the apparent lack of an association of lithium treatment with expected reduction of suicidal behavior, Shuy et al. [13] consider the possibility of the short exposure of lithium and the relative infrequency of suicide acts.

In this review, we focus on the effects of trace lithium, defined as very low levels of lithium far below the therapeutic levels commonly used in routine psychiatric practice. This review is not intended to be a systematic or quantitative analysis but rather a qualitative exploration of the potential effects of trace lithium in the prevention of suicide and dementia prevention. The aim of this review is to reconfirm these effects and to specifically suggest the mechanisms at low lithium levels.

2. Lithium for Suicide Prevention

Wang et al. [14] showed that among 1647 patients with bipolar disorder across 13 Asian countries, the doses of mood stabilizers averaged 584 (confidence interval, 565–603 lithium-equivalent mg/day), and 13.1% of the patients given greater than 900 mg/day were younger, male, currently hospitalized, not currently depressed, and reported lifetime suicidal ideation in comparison to those given lower doses. Probably, higher doses of lithium may be required to treat lifetime suicidal ideation in some patients.

Conversely, since the initial report of a significant inverse association between lithium levels in drinking water and suicide rates across 27 Texas counties [15], numerous studies have investigated the association between lithium levels in drinking water and suicide rates, including our own studies [16–18]. For instance, Ohgami et al. [16] measured lithium levels in drinking water in the Oita prefecture of Japan and calculated the suicide standardized mortality ratio (SMR). Their findings indicated that lithium levels were significantly and inversely associated with suicide SMRs, suggesting that even trace amounts of lithium levels in drinking water may help reduce suicide risk in the general population. In contrast, Kabacs et al. [19] found no significant association between lithium in drinking water and suicide rates in the East of England. Meanwhile, Kapusta et al. [20] reconfirmed a significantly inverse association between lithium levels in drinking water and suicide SMRs in Austria, even after adjusting for socioeconomic factors.

Since these studies, further epidemiological investigations have continued to explore this association. Recently, Kugimiya et al. [21] conducted an extensive study across all 808 cities and wards in Japan (785 cities from 46 prefectures and 23 wards of Tokyo), investigating the association between lithium levels in drinking water and suicide SMRs during the 7 years from 2010 to 2016. Using multiple regression analyses adjusted for the size of each population size and other relevant factors, they found significantly inverse associations of lithium levels with total and male SMRs but not with female SMRs [21]. These findings further strengthen the evidence supporting an inverse association between lithium levels in drinking water and suicide rates, particularly in men. Finally, recent meta-analyses [22–24] have confirmed that higher lithium levels in drinking water may be associated with lower suicide rates.

With regard to the anti-suicidal effects of therapeutic levels of lithium in clinical settings, a recent meta-analysis [25] reported that the pooled suicide rate was 0.2% for individuals randomized to lithium and 0.4% for those receiving a placebo or treatment as usual, which was not a statistically significant difference [odds ratio = 0.42, 95% confidence interval (CI) 0.01–4.5]. However, a recent reanalysis [9] appropriately criticized the methodological flaws in that meta-analysis [25] and revealed a larger effect size. The reanalysis showed that the pooled suicide rate was 0.3% for individuals randomized to lithium and 1.69% for those receiving placebo or treatment as usual, with an odds ratio of 0.25 (95% CI 0.08–0.83), suggesting a 75% reduction in suicide completion associated with lithium drugs in patients with mood disorders.

It is challenging to directly compare the lithium effects of trace lithium levels with therapeutic levels due to differences in the populations studied, namely, the general population versus patients with mood disorders. However, the contrast between a 58% reduction in suicide completion in the general population associated with trace levels of lithium levels [22] and a 75% reduction of suicide completion in patients with mood disorders associated with therapeutic levels of lithium levels [9] suggests that higher levels of lithium may exert stronger anti-suicidal effects. As illustrated in Figure 1, trace lithium levels may predominantly elicit anti-aggressive and anti-impulsive effects, which in turn could contribute to reducing suicide risk. On the other hand, therapeutic levels of lithium levels appear to generate both mood-stabilizing effects and anti-aggressive and anti-impulsive effects, enhancing their overall anti-suicidal impact. This could explain why therapeutic levels of lithium may offer more potent anti-suicidal effects than trace levels of lithium.

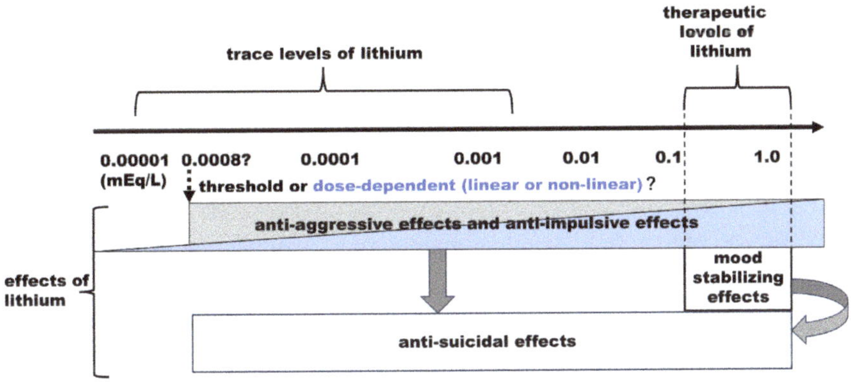

Figure 1. The effects of lithium from trace levels to therapeutic levels. Trace lithium levels primarily produce anti-aggressive/anti-impulsive effects, contributing to anti-suicidal outcomes, while therapeutic levels of lithium add mood-stabilizing effects. The threshold for anti-aggressive/anti-impulsive effects may occur at 0.0008 mEq/L as a threshold, or the effect could follow a dose–response pattern (linear or non-linear).

Epidemiological studies investigating the association between lithium levels in drinking water and suicide rates may be subject to ecological fallacy. To address this, we further examined the association between trace lithium levels and suicide attempts at the individual level [26–28]. These participants were not receiving lithium therapy and were presumed to ingest trace lithium through drinking water and food sources. The most recent study by Izumi et al. [28] explored the effects of naturally absorbed trace lithium alongside unsaturated fatty acids such as eicosapentaenoic acid (EPA), docosahexaenoic acid (DHA), and arachidonic acid (AA) on deliberate self-harm and suicide attempts in 234 patients. The findings indicated that higher serum lithium levels were significantly associated with fewer suicide attempts and fewer deliberate self-harm, even after adjusting for age, sex, and unsaturated fatty acids (EPA, DHA, or AA). Mean serum lithium levels in the suicide attempt group, self-harm group, and control group were 4.04, 4.37, and 5.62 μg/L, respectively, corresponding to 0.00058, 0.00063, and 0.00081 mEq/L [28]. If the association between lithium levels and anti-aggressive/anti-impulsive effects has a cut-off value (a threshold, it may lie around 0.0008 mEq/L (gray section of Figure 1). Alternatively, there may be a dose–response relationship between serum lithium levels (blue section of Figure 1). Currently, it is unclear whether this association follows a threshold or a dose–response pattern.

Although our studies focused on suicide attempters, Ando et al. [29] examined suicide completers by measuring lithium levels in the aqueous humor, which remains relatively stable after death. Their findings revealed that lithium levels in the aqueous humor of 12 suicide completers (mean 0.50 μg/L) were significantly lower than those in 16 controls with non-suicidal deaths (mean 0.92 μg/L). These results suggest that similar to suicide attempters, suicide completers may have significantly lower lithium levels in the body compared to non-suicidal controls. This represents important evidence for the inverse association between lithium levels and suicide completion, as strictly speaking, suicide attempts and suicide completion are different.

One of the animal studies [30] showed that lithium decreased shock-induced aggression in mice. One potential mechanism by which low-dose lithium may prevent suicide prevention is its effect on testosterone levels, which are associated with aggression and impulsivity. Testosterone is the primary androgen hormone responsible for male sexual development and the maintenance of male secondary sexual characteristics. In men, testosterone is predominantly secreted by Leydig cells in the testes, though a smaller amount of testosterone is also produced by the adrenal glands. In women, lesser quantities of testosterone are secreted by the adrenal glands and ovaries [31]. It has been suggested that testosterone modulates social-emotional behavior in men through prefrontal-amygdala functional connectivity [31]. Sher et al. [32,33] also reported that higher testosterone levels were linked with increased suicidality in both men and women, with testosterone levels generally being much higher in men. Furthermore, lithium administration has been shown to reduce testosterone levels, suggesting that lithium may lower suicidality, particularly in men, by decreasing testosterone levels [34]. This points to the possibility that lithium's inhibition of testosterone may reduce aggression and impulsivity, thus contributing to suicide prevention.

With respect to trace lithium, an observational study conducted in China followed 796 college students who were recruited from June 2013 (baseline) to 2014 to investigate the effects of exposure to metal/metalloid elements on male fertility [35]. During both phases, semen and blood samples were collected to assess semen quality and levels of six sex hormones. One key finding was that the median urinary lithium levels (μg/L) were 16.5 at baseline and 16.69 at follow-up, while median serum testosterone levels (nmol/L) were 4.3 at baseline and 4.0 at follow-up. After adjusting for confounding factors, urinary lithium levels were found to be significantly and inversely associated with serum testosterone [35], suggesting that trace lithium may be inversely associated with serum testosterone levels [35]. These results suggest that even trace lithium may also reduce testosterone levels, which in turn could lower aggression and impulsivity, leading to suicide prevention, as illustrated in Figure 2.

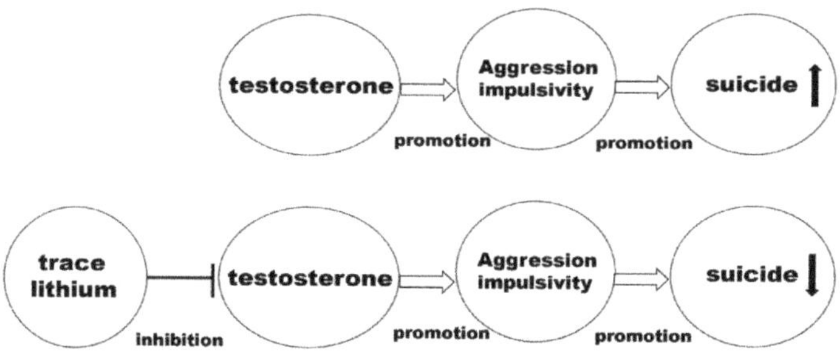

Figure 2. Lithium against testosterone, aggression/impulsivity, and suicide. Trace lithium may inhibit testosterone, which may decrease aggression and impulsivity, leading to suicide prevention.

3. Lithium for Dementia Prevention

Nunes et al. [36] randomized 113 patients with Alzheimer's disease (AD) and Mini Mental State Evaluation (MMSE) scores ranging from 9 to 21 to receive either lithium (300 μg/day; n = 58) or placebo (n = 55) treatments in a 15-month, randomized, placebo-controlled, double-blind trial. The lithium group demonstrated no decline in their performance on the MMSE scores, whereas the placebo group showed a decline, suggesting a potential neuroprotective effect of trace lithium on AD progression [36].

Regarding epidemiological studies, Kessing et al. [37] examined the potential association between lithium levels in drinking water and the incidence of dementia in the general population of Denmark. They found that higher lithium levels in drinking water were nonlinearly associated with a reduced incidence of dementia. In contrast, Parker et al. [38] reported no significant association between higher lithium levels in drinking water and the prevalence of dementia. These conflicting results highlight the variability in epidemiological studies. Additionally, Fajardo et al. [39] investigated AD mortality and demonstrated a significantly inverse association between lithium levels in drinking water and changes in AD mortality, suggesting that trace lithium may influence AD mortality rates.

Recently, Muronaga et al. [40] investigated the association between lithium levels in drinking water and the prevalence of AD across all 808 cities and wards in Japan. Their study demonstrated a significantly inverse association between lithium levels and the prevalence of AD in females but not in males or the overall population after adjusting for several factors associated with dementia. These findings suggest that higher lithium levels in drinking water may be associated with a low prevalence of AD in females, though no such association was observed in males, which contrasts with our previous findings related to suicide rates [21].

With regard to the mechanisms of low-dose lithium for dementia prevention, several factors have been proposed in preclinical studies [41]. For example, as shown in Figure 3A, low-dose lithium may inhibit Glycogen Synthase Kinase-3β (GSK-3β) [42,43] and thereby decrease amyloidβ_{42} levels [42] and senile plaques [43,44]. Also, low-dose lithium may inhibit pro-inflammatory cytokine (IL-6, IL-8, etc.) [45,46] and neuroinflammation. Moreover, as shown in Figure 3B, low-dose lithium may promote Brain-derived Neurotrophic Factor (BDNF) [44,47] and neurogenesis [42,44]. In contrast to the mechanism of lithium for suicide prevention, there are many acting sites of lithium for dementia prevention, although there may be many unrevealed acting sites of lithium for suicide prevention.

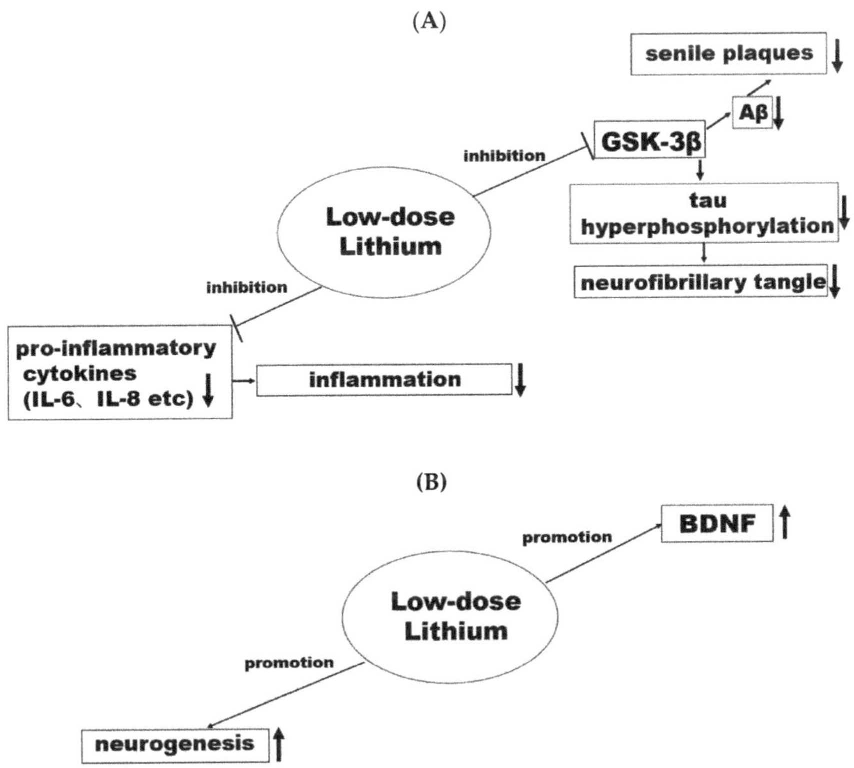

Figure 3. The mechanisms of trace lithium for dementia prevention. (**A**): lithium inhibition, (**B**): lithium promotion. Low-lithium may inhibit Glycogen Synthase Kinase-3β (GSK-3β) and pro-inflammatory cytokine (IL-6, IL-8, etc.) and promote Brain-derived Neurotrophic Factor (BDNF) and neurogenesis.

4. New Lithium Product

Nanolithium is an experimental product utilizing a novel drug-delivery technology (Aonys®, MEDESIS PHARMA, Baillargues, France) that optimizes its bioavailability while minimizing its toxicity profile [48].

The therapeutic doses of lithium used in Nanolithium are more than 50 times lower than the minimal doses used in traditional lithium salts. GSK-3β inhibition is believed to be central to Nanolithium's mechanism of action, leading to a reduction in the production of toxic amyloid plaques and a decrease in tau hyperphosphorylation, which may potentially improve both neuropsychiatric symptoms and cognitive decline. Currently, several clinical studies are underway [48].

5. Discussion

Post and Rybakowski [49] depicted the multiple assets of lithium beyond its anti-manic effects, which were demonstrated in recent years as follows.

(1) Diagnosis: Lithium prevents unipolar and bipolar depression.
(2) Suicide: Lithium prevents suicides in patients and in the general population. Lithium prevents suicide in the general population at minuscule doses in the water supply.
(3) Augmentation: Lithium enhances the effects of other mood stabilizers and atypical antipsychotics.
(4) Hippocampus: Lithium increases the volume of the hippocampus.
(5) Gray/white matter: Lithium prevents or reverses cortical gray matter and white matter tract deficits.
(6) Circadian rhythm: Lithium normalizes circadian rhythm abnormalities and has immune-modulating effects.
(7) Immune system: Lithium has immune-modulating effects.
(8) Telomeres: Lithium increases the length of telomeres shortened by episodes, stress, and aging.
(9) Memory: Lithium prevents or slows memory deterioration in mild cognitive impairment over one year.
(10) Dementia: Lithium reduces the incidence of a diagnosis of dementia in old age.
(11) Life expectancy: Lithium prolongs life expectancy and reduces the incidence of all-cause mortality.
(12) All-cause mortality: Lithium reduces the incidence of all-cause mortality.

Among these effects, "Lithium prevents suicide in the general population at minuscule doses in the water supply." and "Lithium prevents or slows memory deterioration in mild cognitive impairment over one year. Lithium reduces the incidence of a diagnosis of dementia in old age." match the present findings.

The present findings suggest that trace lithium may be associated with lower suicide rates and reduced dementia rates, although most studies have been cross-sectional, and causality cannot be definitively established. Nonetheless, the potential value of trace lithium appears significant, as there is currently no definitive treatment for suicide prevention or dementia, and trace lithium is likely to avoid the side effects associated with so-called therapeutic lithium levels.

Regarding the experimental drug containing trace lithium named Nanolithium [48], it is expected that the ongoing randomized, placebo-controlled, parallel-group studies can clarify its effects on Alzheimer's disease. Additionally, it is plausible that Nanolithium may reduce suicide attempts and completions, warranting further investigation. Notably, in Japan, one psychiatrist with extensive experience has prescribed 1 to 2 mg of lithium alongside powdered stomach medication for young psychiatric patients [50]. While the possibility of a placebo effect cannot be ruled out, this approach represents a much simpler treatment compared to Nanolithium.

With regard to the role of trace lithium as an essential trace element [51,52], essential trace elements are generally defined as dietary minerals required in very minute quantities for the proper growth, development, and physiological functioning of an organism. [50]. Regrettably, the available data on trace lithium's role in human growth, development, and physiology are limited. Nonetheless, a deficiency of trace lithium may contribute to difficulties in controlling aggression and impulsivity, potentially leading to self-harm and/or suicide. Additionally, in line with Schrauzer and Shrestha [15], Kohno et al. [53] demonstrated that lithium levels in drinking water were significantly and inversely associated with crime rates, suggesting that trace lithium may influence aggressiveness and impulsivity toward others as well as oneself.

6. Limitations

This is a qualitative and non-systematic review. As such, the references were deviated to the authors' preference, although we tried to collect references as much as fairly. Also, the quality of the references was various due to different sample sizes and methodologies.

7. Conclusions

The present findings suggest that higher trace lithium levels may be associated with lower suicide rates and reduced dementia rates. Probably, trace lithium may inhibit testosterone and thereby mitigate aggression and impulsivity and decrease suicide. Also, trace lithium may inhibit GSK-3β and thereby lower amyloid β and tau hyperphosphorylation, and inhibit pro-inflammatory cytokines such as IL 6 and IL 8 and thereby mitigate inflammation, whereas trace lithium may promote BDNF and neurogenesis in the general population.

8. Future Directions

If trace lithium levels are confirmed to be effective for suicide prevention and dementia prevention, it is necessary to clarify whether the association between lithium levels and its anti-suicidal effects and anti-dementia effects follows a dose–response pattern—whether it is linear, non-linear, or characterized by a threshold effect. Eventually, prospective and multicenter studies are warranted to confirm that trace lithium can be effective for suicide prevention and for dementia prevention. Also, further research on pharmacogenetic testing [54] is required for suicide prevention and dementia prevention.

Author Contributions: Conceptualization, T.T.; writing—original draft preparation, T.T.; writing—review and editing, H.H., M.M., T.I. and K.K. All authors have read and agreed to the published version of the manuscript.

Funding: This research received no external funding.

Conflicts of Interest: The authors declare no conflicts of interest.

References

1. Cade, J.F. Lithium salts in the treatment of psychotic excitement. *Med. J. Aust.* **1949**, *2*, 349–352. [CrossRef] [PubMed]
2. Hong, Y.; Huang, W.; Cao, D.; Xu, J.; Wei, H.; Zhang, J.; Wang, L. A cumulative Bayesian network meta-analysis on the comparative efficacy of pharmacotherapies for mania over the last 40 years. *Psychopharmacology* **2022**, *239*, 3367–3375. [CrossRef] [PubMed]
3. Hsu, C.W.; Tsai, S.Y.; Tseng, P.T.; Liang, C.S.; Vieta, E.; Carvalho, A.F.; Stubbs, B.; Kao, H.Y.; Tu, Y.K.; Lin, P.Y. Differences in the prophylactic effect of serum lithium levels on depression and mania in bipolar disorder: A dose-response meta-analysis. *Eur. Neuropsychopharmacol.* **2022**, *58*, 20–29. [CrossRef] [PubMed]
4. Kishi, T.; Ikuta, T.; Matsuda, Y.; Sakuma, K.; Okuya, M.; Nomura, I.; Hatano, M.; Iwata, N. Pharmacological treatment for bipolar mania: A systematic review and network meta-analysis of double-blind randomized controlled trials. *Mol. Psychiatry* **2022**, *27*, 1136–1144. [CrossRef]
5. Nuñez, N.A.; Joseph, B.; Pahwa, M.; Kumar, R.; Resendez, M.G.; Prokop, L.J.; Veldic, M.; Seshadri, A.; Biernacka, J.M.; Frye, M.A.; et al. Augmentation strategies for treatment resistant major depression: A systematic review and network meta-analysis. *J. Affect. Disord.* **2022**, *302*, 385–400. [CrossRef]
6. Zhou, X.; Ravindran, A.V.; Qin, B.; Del Giovane, C.; Li, Q.; Bauer, M.; Liu, Y.; Fang, Y.; da Silva, T.; Zhang, Y.; et al. Comparative efficacy, acceptability, and tolerability of augmentation agents in treatment-resistant depression: Systematic review and network meta-analysis. *J. Clin. Psychiatry* **2015**, *76*, e487–e498. [CrossRef]

7. Nelson, J.C.; Baumann, P.; Delucchi, K.; Joffe, R.; Katona, C. A systematic review and meta-analysis of lithium augmentation of tricyclic and second generation antidepressants in major depression. *J. Affect. Disord.* **2014**, *168*, 269–275. [CrossRef]
8. Wilkinson, S.T.; Trujillo Diaz, D.; Rupp, Z.W.; Kidambi, A.; Ramirez, K.L.; Flores, J.M.; Avila-Quintero, V.J.; Rhee, T.G.; Olfson, M.; Bloch, M.H. Pharmacological and somatic treatment effects on suicide in adults: A systematic review and meta-analysis. *Depress. Anxiety* **2022**, *39*, 100–112. [CrossRef]
9. Ghaemi, S.N. The pseudoscience of lithium and suicide: Reanalysis of a misleading meta-analysis. *J. Psychopharmacol.* **2024**, *38*, 597–603. [CrossRef]
10. Lu, Q.; Lv, H.; Liu, X.; Zang, L.; Zhang, Y.; Meng, Q. Lithium therapy's potential to lower dementia risk and the prevalence of Alzheimer's disease: A meta-analysis. *Eur. Neurol.* **2024**, *87*, 93–104. [CrossRef]
11. Velosa, J.; Delgado, A.; Finger, E.; Berk, M.; Kapczinski, F.; de Azevedo Cardoso, T. Risk of dementia in bipolar disorder and the interplay of lithium: A systematic review and meta-analyses. *Acta Psychiatr. Scand.* **2020**, *141*, 510–521. [CrossRef] [PubMed]
12. Matsunaga, S.; Kishi, T.; Annas, P.; Basun, H.; Hampel, H.; Iwata, N. Lithium as a treatment for Alzheimer's disease: A systematic review and meta-analysis. *J. Alzheimers Dis.* **2015**, *48*, 403–410. [CrossRef] [PubMed]
13. Shuy, Y.K.; Santharan, S.; Chew, Q.H.; Lin, S.-K.; Ouyang, W.-C.; Chen, C.-K.; Park, S.-C.; Jang, O.-J.; Park, J.H.; Chee, K.-Y.; et al. Pharmacoepidemiology and clinical correlates of lithium treatment for bipolar disorder in Asia. *J. Clin. Psychopharmacol.* **2024**, *44*, 117–123. [CrossRef] [PubMed]
14. Wang, Y.; Chew, Q.H.; Lin, S.-K.; Yang, S.-Y.; Ouyang, W.-C.; Chen, C.-K.; Park, S.-C.; Jang, O.-J.; Park, J.H.; Chee, K.-Y.; et al. Dosing of mood stabilizers for bipolar disorder patients in the research on Asian psychotropic prescription patterns consortium study. *J. Clin. Psychopharmacol.* **2022**, *42*, 293–297. [CrossRef]
15. Schrauzer, G.N.; Shrestha, K.P. Lithium in drinking water and the incidences of crimes, suicides, and arrests related to drug addictions. *Biol. Trace Elem. Res.* **1990**, *25*, 105–113. [CrossRef]
16. Ohgami, H.; Terao, T.; Shiotsuki, I.; Ishii, N.; Iwata, N. Lithium levels in drinking water and risk of suicide. *Br. J. Psychiatry* **2009**, *194*, 464–465; discussion 446. [CrossRef]
17. Ishii, N.; Terao, T.; Araki, Y.; Kohno, K.; Mizokami, Y.; Shiotsuki, I.; Hatano, K.; Makino, M.; Kodama, K.; Iwata, N. Low risk of male suicide and lithium in drinking water. *J. Clin. Psychiatry* **2015**, *76*, 319–326. [CrossRef]
18. Shiotsuki, I.; Terao, T.; Ishii, N.; Takeuchi, S.; Kuroda, Y.; Kohno, K.; Mizokami, Y.; Hatano, K.; Tanabe, S.; Kanehisa, M.; et al. Trace lithium is inversely associated with male suicide after adjustment of climatic factors. *J. Affect. Disord.* **2016**, *189*, 282–286. [CrossRef]
19. Kabacs, N.; Memon, A.; Obinwa, T.; Stochl, J.; Perez, J. Lithium in drinking water and suicide rates across the East of England. *Br. J. Psychiatry* **2011**, *198*, 406–407. [CrossRef]
20. Kapusta, N.D.; Mossaheb, N.; Etzersdorfer, E.; Hlavin, G.; Thau, K.; Willeit, M.; Praschak-Rieder, N.; Sonneck, G.; Leithner-Dziubas, K. Lithium in drinking water and suicide mortality. *Br. J. Psychiatry* **2011**, *198*, 346–350. [CrossRef]
21. Kugimiya, T.; Ishii, N.; Kohno, K.; Kanehisa, M.; Hatano, K.; Hirakawa, H.; Terao, T. Lithium in drinking water and suicide prevention: The largest nationwide epidemiological study from Japan. *Bipolar Disord.* **2021**, *23*, 33–40. [CrossRef] [PubMed]
22. Barjasteh-Askari, F.; Davoudi, M.; Amini, H.; Ghorbani, M.; Yaseri, M.; Yunesian, M.; Mahvi, A.H.; Lester, D. Relationship between suicide mortality and lithium in drinking water: A systematic review and meta-analysis. *J. Affect. Disord.* **2020**, *264*, 234–241. [CrossRef] [PubMed]
23. Memon, A.; Rogers, I.; Fitzsimmons, S.M.D.D.; Carter, B.; Strawbridge, R.; Hidalgo-Mazzei, D.; Young, A.H. Association between naturally occurring lithium in drinking water and suicide rates: Systematic review and meta-analysis of ecological studies. *Br. J. Psychiatry* **2020**, *217*, 667–678. [CrossRef] [PubMed]
24. Eyre-Watt, B.; Mahendran, E.; Suetani, S.; Firth, J.; Kisely, S.; Siskind, D. The association between lithium in drinking water and neuropsychiatric outcomes: A systematic review and meta-analysis from across 2678 regions containing 113 million people. *Aust. N. Z. J. Psychiatry* **2021**, *55*, 139–152. [CrossRef]
25. Nabi, Z.; Stansfeld, J.; Plöderl, M.; Wood, L.; Moncrieff, J. Effects of lithium on suicide and suicidal behaviour: A systematic review and meta-analysis of randomised trials. *Epidemiol. Psychiatr. Sci.* **2022**, *31*, e65. [CrossRef]
26. Kanehisa, M.; Terao, T.; Shiotsuki, I.; Kurosawa, K.; Takenaka, R.; Sakamoto, T.; Shigemitsu, O.; Ishii, N.; Hatano, K.; Hirakawa, H. Serum lithium levels and suicide attempts: A case-controlled comparison in lithium therapy-naive individuals. *Psychopharmacology* **2017**, *234*, 3335–3342. [CrossRef]
27. Kurosawa, K.; Terao, T.; Kanehisa, M.; Shiotsuki, I.; Ishii, N.; Takenaka, R.; Sakamoto, T.; Matsukawa, T.; Yokoyama, K.; Ando, S.; et al. Naturally absorbed polyunsaturated fatty acids, lithium, and suicide-related behaviors: A case-controlled study. *J. Affect. Disord.* **2018**, *241*, 200–205. [CrossRef]
28. Izumi, T.; Kanehisa, M.; Terao, T.; Shiotsuki, I.; Shirahama, M.; Satoh, M.; Muronaga, M.; Kohno, K.; Hirakawa, H.; Etoh, M.; et al. Naturally absorbed lithium may prevent suicide attempts and deliberate self-harm while eicosapentaenoic acid may prevent deliberate self-harm and arachidonic acid may be a risk factor for deliberate self-harm: The updated different findings in new analyses. *Front. Psychiatry* **2022**, *13*, 1083739. [CrossRef]
29. Ando, S.; Suzuki, H.; Matsukawa, T.; Usami, S.; Muramatsu, H.; Fukunaga, T.; Yokoyama, K.; Okazaki, Y.; Nishida, A. Comparison of lithium levels between suicide and non-suicide fatalities: Cross-sectional study. *Transl. Psychiatry* **2022**, *12*, 466. [CrossRef]
30. Kovacsics, C.E.; Gould, T.D. Shock-induced aggression in mice is modified by lithium. *Pharmacol. Biochem. Behav.* **2010**, *94*, 380–386. [CrossRef]

31. Sher, L. Testosterone and suicidal behavior in bipolar disorder. *Int. J. Environ. Res. Public Health* **2023**, *20*, 2502. [CrossRef]
32. Sher, L.; Grunebaum, M.F.; Sullivan, G.M.; Burke, A.K.; Cooper, T.B.; Mann, J.J.; Oquendo, M.A. Testosterone levels in suicide attempters with bipolar disorder. *J. Psychiatr. Res.* **2012**, *46*, 1267–1271. [CrossRef]
33. Sher, L.; Grunebaum, M.F.; Sullivan, G.M.; Burke, A.K.; Cooper, T.B.; Mann, J.J.; Oquendo, M.A. Association of testosterone levels and future suicide attempts in females with bipolar disorder. *J. Affect. Disord.* **2014**, *166*, 98–102. [CrossRef]
34. Sher, L. Suicide in men. *J. Clin. Psychiatry* **2015**, *76*, e371–e372. [CrossRef]
35. Chai, Z.; Zhang, G.; Ling, X.; Dong, T.; Wang, J.; Zhang, Y.; Zou, P.; Yang, H.; Zhou, N.; Chen, Q.; et al. Low-level and combined exposure to environmental metal elements affects male reproductive outcomes: Prospective MARHCS study in population of college students in Chongqing, China. *Sci. Total Environ.* **2022**, *828*, 154395. [CrossRef]
36. Nunes, M.A.; Viel, T.A.; Buck, H.S. Microdose lithium treatment stabilized cognitive impairment in patients with Alzheimer's disease. *Curr. Alzheimer Res.* **2013**, *10*, 104–107. [CrossRef]
37. Kessing, L.V.; Gerds, T.A.; Knudsen, N.N.; Jørgensen, L.F.; Kristiansen, S.M.; Voutchkova, D.; Ernstsen, V.; Schullehner, J.; Hansen, B.; Andersen, P.K.; et al. Association of lithium in drinking water with the incidence of dementia. *JAMA Psychiatry* **2017**, *74*, 1005–1010. [CrossRef]
38. Parker, W.F.; Gorges, R.J.; Gao, Y.N.; Zhang, Y.; Hur, K.; Gibbons, R.D. Association between groundwater lithium and the diagnosis of bipolar disorder and dementia in the United States. *JAMA Psychiatry* **2018**, *75*, 751–754. [CrossRef]
39. Fajardo, V.A.; Fajardo, V.A.; LeBlanc, P.J.; MacPherson, R.E.K. Examining the relationship between trace lithium in drinking water and the rising rates of age-adjusted Alzheimer's disease mortality in Texas. *J. Alzheimers Dis.* **2018**, *61*, 425–434. [CrossRef]
40. Muronaga, M.; Terao, T.; Kohno, K.; Hirakawa, H.; Izumi, T.; Etoh, M. Lithium in drinking water and Alzheimer's dementia: Epidemiological Findings from National Data Base of Japan. *Bipolar Disord.* **2022**, *24*, 788–794. [CrossRef]
41. Manchia, M.; Paribello, P.; Pinna, M.; Steardo, L.; Carpiniello, B.; Pinna, F.; Pisanu, C.; Squassina, A.; Hajek, T. Lithium and its effects: Does dose matter? *Int. J. Bipolar Disord.* **2024**, *12*, 23. [CrossRef]
42. Wilson, E.N.; Do Carmo, S.; Iulita, M.F.; Hall, H.; Ducatenzeiler, A.; Marks, A.R.; Allard, S.; Jia, D.T.; Windheim, J.; Cuello, A.C. BACE1 inhibition by microdose lithium formulation NP03 rescues memory loss and early stage amyloid neuropathology. *Transl. Psychiatry* **2017**, *7*, e1190. [CrossRef]
43. Pereira, A.A.R.; Pinto, A.M.; Malerba, H.N.; Toricelli, M.; Buck, H.S.; Viel, T.A. Microdose lithium improves behavioral deficits and modulates molecular mechanisms of memory formation in female SAMP-8, a mouse model of accelerated aging. *PLoS ONE* **2024**, *19*, e0299534. [CrossRef]
44. Nunes, M.A.; Schöwe, N.M.; Monteiro-Silva, K.C.; Baraldi-Tornisielo, T.; Souza, S.I.G.; Balthazar, J.; Albuquerque, M.S.; Caetano, A.L.; Viel, T.A.; Buck, H.S. Chronic microdose lithium treatment prevented memory loss and neurohistopathological changes in a transgenic mouse model of Alzheimer's disease. *PLoS ONE* **2015**, *10*, e0142267. [CrossRef]
45. Toricelli, M.; Evangelista, S.R.; Buck, H.S.; Viel, T.A. Microdose lithium treatment reduced inflammatory factors and neurodegeneration in organotypic hippocampal culture of old SAMP-8 mice. *Cell Mol. Neurobiol.* **2021**, *41*, 1509–1520. [CrossRef]
46. Viel, T.; Chinta, S.; Rane, A.; Chamoli, M.; Buck, H.; Andersen, J. Microdose lithium reduces cellular senescence in human astrocytes: A potential pharmacotherapy for COVID-19? *Aging* **2020**, *12*, 10035–10040. [CrossRef]
47. De-Paula, V.J.; Gattaz, W.F.; Forlenza, O.V. Long-term lithium treatment increases intracellular and extracellular brain-derived neurotrophic factor (BDNF) in cortical and hippocampal neurons at subtherapeutic concentrations. *Bipolar Disord.* **2016**, *18*, 692–695. [CrossRef]
48. Guilliot, S.; Wilson, E.N.; Touchon, J.; Soto, M.E. Nanolithium, a new treatment approach to Alzheimer's disease: A review of existing evidence and clinical perspectives. *J. Prev. Alzheimers Dis.* **2024**, *11*, 428–434. [CrossRef]
49. Post, R.M.; Rybakowski, J.K. What patients with bipolar disorder need to know about lithium. *Pharmaceuticals* **2024**, *17*, 1223. [CrossRef]
50. Sugimoto, T. The practical experience of prescribing very low doses of lithium and psychotropics. In Proceedings of the 120th Annual Meeting of the Japanese Society of Psychiatry and Neurology, Hokkaido, Japan, 20–22 June 2024; Volume S94–4, p. 636. (In Japanese)
51. Terao, T. Is lithium potentially a trace element? *World J. Psychiatry* **2015**, *5*, 1–3. [CrossRef]
52. Ishii, N.; Terao, T. Trace lithium and mental health. *J. Neural. Transm.* **2018**, *125*, 223–227. [CrossRef]
53. Kohno, K.; Ishii, N.; Hirakawa, H.; Terao, T. Lithium in drinking water and crime rates in Japan: Cross-sectional study. *BJPsych Open* **2020**, *6*, e122. [CrossRef]
54. Adiukwu, F.; Adesokun, O.; Essien, E.; Yalcin, N.; Ransing, R.; Nagendrappa, S.; Jatchavala, C.; Olakunke, A.B.; Nawaz, F.A.; Khan, N.A. Pharmacogenetic Testing in Psychiatry: Perspective on Clinical Utility. *Asian J. Psychiatry* **2023**, *86*, 103674. [CrossRef]

pharmaceuticals

MDPI

Review

At 75 Years, Is Lithium the Gold Standard in Bipolar Disorder?

Robert H. Belmaker

Department of Psychiatry, Faculty of Health Sciences, Ben Gurion University of the Negev, Beersheva 8410501, Israel; belmaker@bgu.ac.il

Abstract: Lithium was the first discovered mood stabilizer for bipolar disorder. However, several antiepileptics such as valproate and carbamazepine have also proven to be mood stabilizers in all phases of bipolar disorder with profiles barely distinguishable from lithium. Second-generation antipsychotics were slowly recognized in almost all cases to be useful in all phases of bipolar disorder. These different treatments have not been found to have any common mechanism of action. Indeed, lithium itself has numerous biochemical effects that are often not replicable. The newest biochemical effects described for lithium are exciting but are still merely promising and should not be taught to medical students as facts, especially since they do not explain the fact that several antiepileptic and second-generation antipsychotics also work well in all phases of bipolar disorder. Lithium is not specific to bipolar disorder but has uses in several other psychiatric and nonpsychiatric conditions. Lithium, like most treatments in medicine, is nonspecific; it is still useful in many cases, but is not in any sense a "gold standard".

Keywords: lithium; bipolar disorder; gold standard; lithium response; mechanism of action; second-generation antipsychotics; antiepileptics

1. Introduction

75 years since John Cade's foundational work on lithium in psychiatry is a proper time for celebration, reflection and reconsideration. Indeed, all of psychopharmacology is going through such a historical self-evaluation. Ed Shorter's recent book *The Rise and Fall of the Age of Psychopharmacology* [1] describes the difficulties in industry-sponsored pharmaceutical company trials and the disappointment in the lack of progress in new antipsychotic and new antidepressant compounds. In the lithium area, Tom Ban's scholarly *History of Lithium Research* [2] devotes great effort to ascribing credit for various steps along the way to lithium's clinical use today and less on the actual current place of lithium in a psychopharmacological algorithm. Walter Brown's best-selling book *Lithium: A Doctor, a Drug and a Breakthrough* [3] novelized the story of John Cade and made him into a hero without critical analysis of the fact that Cade himself refused to use lithium for most of his professional life after his early discoveries. Some recent journal articles have celebrated lithium as the "first choice" or "gold standard" for treatment of bipolar disorder [4]. However, other articles have emphasized the increasing concerns of long-term lithium treatment on the kidney [5,6] and on the parathyroid gland, even more than on the thyroid. Many articles have noted that in every major country where lithium has been widely used, lithium's use is declining both on a per capita basis and as a percentage of total psychopharmaceutical medications sold. For some researchers, this is seen as a paradox: Why should a "gold-standard" treatment be facing declining use? However, such phenomena are the rule rather than the exception in medicine. The great impact of the first successful human heart transplant by Christian Bernard in 1967 has not led to

increasing rates of heart transplants throughout the world and certainly no one would consider that good medicine implies that heart transplants should be on the rise every year. The impact of Christian Bernard's successful heart transplant was the fact that it made waves in transplant research in many other disciplines, stimulated the development of medicines to combat transplant rejection, and inspired continuing work in the development of non-biological artificial hearts. Similarly, the discovery of the first antibiotic sulfa drug was of huge importance but no one would measure that importance by the rates of use of sulfa drugs today. No one would call for sulfa to replace other antibiotics. Indeed, no one would expect that all antibiotics have the same mechanism. The discovery of sulfa as an antibiotic by Gerhard Domagk in 1932 was a culmination of Ehrlich's idea that there could be a "magic bullet" to cure a disease and it did not imply that the first magic bullet would the only magic bullet or there would be only one kind of magic bullet.

I would like to participate in the celebration of 75 years since John Cade's discovery and divide my discussion into the following sections:

1 Introduction.
2 Were early studies of lithium's action definitive?
3 What is the meaning of "gold standard", "first line", or "drug of choice"?
4 Is lithium superior to the anticonvulsant mood stabilizers valproate and carbamazepine?
5 Is lithium superior to second-generation antipsychotics as a mood stabilizer?
6 Were first-generation D-2 blockers also mood stabilizers?
7 Do we have a mechanism of action of lithium worth teaching to medical students or residents?
8 Does response to lithium define bipolar disorder and is bipolar disorder always an indication for use of lithium?
9 Lithium and Suicidality.
10 The Dangers of the Intellectual Cult Around Lithium and the Importance of an Empirical Patient-Centered Therapeutic Armamentarium

2. Were Early Studies of Lithium's Action Definitive?

Cade's first report of lithium's usefulness in acute mania was a series of about a dozen cases. His insight into the possible utility of lithium in mania was based on theories of the pathophysiology of mania, uric acid, and the solubility of lithium urate: all speculations now known to be completely false. He had performed some small rodent studies finding a sedative effect of acute lithium injection. However, chronic oral lithium treatment in rodents is not considered to be sedative and acute peritoneal lithium as Cade gave causes painful peritoneal ulceration that certainly reduces rodent activity level but is no indication of psychoactive effect. Cade's early discovery could therefore be considered a comedy of errors rather than an inspiring example of a rational-based discovery. His publication was certainly a brave act. It was a thrust at a paradigm change in world psychiatry which was on the cusp of emerging from psychoanalytic dominance. The story of Cade's discovery cannot be told without the fact that he did not go on to use lithium and actually forbade the use of lithium in his hospital after the original discovery. Lithium after Cade became a major medication in the treatment of congestive heart failure by cardiologists, who after WWII had just begun to understand the role of sodium accumulation in the body as the mechanism of pulmonary edema and congestive heart failure. Lithium chloride tastes salty and was recommended by cardiologists for congestive heart failure as a replacement for sodium chloride in the diet. A wave of fatalities due to lithium intoxication spread around the world and lithium use was condemned worldwide as a sodium substitute in congestive heart failure. Lithium might have been forgotten in psychiatry as well if not for the work of Mogens Schou in Denmark and Samuel Gershon in the United States. Mogens Schou began his lithium studies perhaps based on Cade's report but historians suggest that he may have been even more familiar with work on lithium in affective disorder performed by Danish psychiatrists in the early 1900s and published only in Danish. As often occurs in science, the technology led the way. The development of flame spectrometry allowed the easy measuring of lithium levels during lithium treatment. Schou started his work with studies in dogs and found that lithium could be administered safely if blood levels were monitored and kept within an empirically determined range. This allowed him to begin clinical trials of lithium and he published his seminal controlled clinical trial using the "mirror method" in 1970. This study, for which Schou received much highly deserved historical credit, was not accepted immediately by everyone in the field. As an example, Barry Blackwell, a distinguished psychiatrist, published a blistering critique of Mogen Schou's study and methodology which he found unconvincing. British and American use of lithium in bipolar disorder was delayed by these critics. Sam Gershon had been a student of Cade's in Australia and upon moving to the United States began studies of lithium with the aim of understanding its biochemical effects on the brain. He conceived the idea that lithium as a simple ion might give us unique insights into the pathophysiology of the brain in psychiatric disorder. His translational approach assumed rather than studied the actual clinical usefulness of lithium. Indeed, when the discoveries of valproate and carbamazepine as mood stabilizers occurred, several critical reviews were published claiming that lithium was of unproven effectiveness in bipolar disorder. In retrospect, the studies performed in the 1960s and 70s did not use the same rating scales as are used today, did not have the same level of blindness as used today, and did not have the large numbers of patients used in pharmaceutical research today. While lithium clearly has the longest history of use for bipolar disorder, it is hard to know how to include or exclude the early studies from meta-analyses of lithium's efficacy.

3. What Is the Meaning of "Gold Standard", "First Line", or "Drug of Choice"?

The "gold-standard" treatment in medicine is the best available treatment at a given time. In most areas of medicine this is a very dynamic parameter and the "gold standard" of treatment changes as newer and better treatments become available. The "gold standard" certainly shouldn't be the "latest treatment" because often as drugs lose their patented status, pharmaceutical companies synthesize "me-too" drugs that are very similar to a previous "gold standard" but with no proven additional efficacy. These "me-too" drugs are more profitable than the drug that has become generic and are highly promoted by advertising. The Food and Drug Administration (FDA) demands that such a "me-too" compound be shown to be more efficacious than placebo in at least two large clinical trials but does not demand before registration that the drug be shown to be better than the previous "gold standard". Therefore, in the eyes of many clinicians, the latest, most expensive drug becomes the "gold standard". This is emphatically not true. However, in other areas of medicine the word "gold standard" can sometimes be confused with the concept of the "classic drug". For instance, digitalis could be said by some to be the "gold standard" for the treatment of congestive heart failure. It certainly was discovered very early on in the history of medicine, is somewhat unique in being a natural product found in the leaves of the foxglove plant, and has also been highly heuristic because of its specific action on the sodium pump in the heart cells which led to discovery of the ion channels and natural regulation of heart function. However, digitalis itself is rarely used nowadays and has been supplanted in clinical use by other compounds which have much wider therapeutic indices, less toxicity, and in some cases greater effectiveness. Digitalis is the classic cardiac glycoside, the classic treatment of congestive heart failure, but not the "gold standard" for every patient today with a diagnosis of congestive heart failure.

Another concept in medicine different from the concept of gold standard is the concept of a "first-line treatment". A first-line treatment is a treatment which a physician may choose for treatment of a specific disorder in a newly diagnosed patient who has not failed in any other treatment. In most diagnoses, there are many first-line treatments. For instance, hypertension may be treated by any number of drugs with any number of different mechanisms of action: for instance, (a) a calcium-channel blocker, of which several are available; (b) an angiotensin-converting enzyme inhibitor (ACE), any number of which are available; (c) an angiotensin receptor blocker, any number of which are available; (d) a thiazide diuretic, any number of which are available; and (e) an alpha receptor antagonist, any number of which are available. All of these drugs may be used as first-line monotherapy. Very often in hypertension treatment, monotherapy is insufficiently effective and combinations of any of the "first-line treatments" represent acceptable "second-line treatment". For cases refractory to such combinations, there are approved drugs that are given as second line, only after the failure of first-line drugs, but they have more side effect risks, more complex pharmacokinetic interactions, less favorable pharmacodynamics (such as the need to be given several times a day), or in some countries for legitimate reasons are more expensive. In many areas of medicine, such as metastatic or recurrent cancer, there are also third-line drugs that are only given to patients who have failed with first- and second-line treatment.

By contrast with the concept of "gold standard" or "first line", the concept of "drug of choice" is less well defined. Often when doctors talk about the "drug of choice", they might be basing their view on what might be best not for all patients with a given diagnosis but for a very specific patient with a unique combination of diagnosis, comorbid disorders, coexisting drug treatment, or economic possibilities. Clearly, the concepts of "gold standard", "first line", and "drug of choice" should not be confused with the concepts of (a) "what everyone is asking about", (b) what makes me as a doctor feel good as a medical professional, (c) what I learned about in medical school, or (d) what my own area of research is.

In psychiatry, Electroconvulsive Therapy (ECT) is sometimes called the "gold standard" of antidepressant treatment. This chapter is not a review of ECT but every fair clinical scientist would agree that ECT is very rarely the "first line" of antidepressant treatment, if depression is defined in the same way as in epidemiological studies that show an incidence of about 20% of the population. ECT might be "first line" in the relatively rare disease of melancholia or acute psychotic depression. ECT, while very effective in the rare disease of melancholia in the short term, is far from a practical long term prophylactic treatment of most depressions that are recurring. It would certainly not be a "gold standard" in the treatment of a chronic resistant depression because the response rate in such patients is low, even if it might be a reasonable third- or fourth-line treatment for such patients. ECT, as almost everyone would agree, is not a "drug of choice" for most depression, even though it is on rare occasions necessary and effective and the right treatment to give, for acute and pervasive suicidality, for instance. Given these many nuances that are so important in medicine, this author has been disturbed by a recent academic crusade to call lithium "more than first line" in bipolar disorder, which sounds like gold standard [4].

Much of the use of the term "gold standard" in relation to lithium is performed by basic scientists who are trying to understand the biochemical mechanism of lithium as a simple ion on the same column of the periodic table as sodium and potassium and find it tantalizing to study in many biochemical systems (see Section 7 of this chapter). These biochemical studies, in order to introduce the reason for the study of lithium biochemistry, call lithium the "gold standard" for treatment of bipolar disorder. However, this large body of biochemical research, much of which has not been replicated and none of which has led to a clear proven mechanism of action of lithium, also cannot be taken in any sense as proof of lithium's efficacy in bipolar disorder.

The soaring prices of gold in the international economic market after the Donald Trump administration has greatly weakened international belief in the dollar as a reserve currency, but has nothing to do with the chemistry of gold as a mineral element and everything to do with mankind's psychological evaluation of the future of economics. The "gold standard" as a phrase teaches its own lesson: USA President Roosevelt took the dollar off the "gold standard" to a large degree in 1933 and USA President Nixon totally eliminated the connection of the dollar to gold in 1971. Both actions were economically rational, based on good economical science, and obviously successful to the patient if the patient is considered both the USA and the world economy.

Lithium is a classic treatment of bipolar disorder and is one of several first-line treatment choices available to the discerning physician based on past history of patient response, patient compliance to blood testing, patient kidney health and general health, patient side effect preference, availability of lab testing, and family history of lithium response. It is not a "gold standard". We have come a long way in 75 years since lithium was the only option for bipolar patients [7].

4. Is Lithium Superior to the Anticonvulsant Mood Stabilizers Valproate and Carbamazepine?

Carbamazepine's potential use in bipolar disorder was discovered by Okuma in Japan and published as a controlled study in Japanese in 1973 [8]. The discovery seems to have been serendipitous and the hypothesis of using carbamazepine, which was a well-known anticonvulsant for bipolar disorder, is not clear. I discussed this with Okuma himself at the 1982 conference in Munich on bipolar disorders, but the answer was not clear to me. Okuma was honored with the International College of Neuropsychopharmacology (CINP) Founders Award for his discovery of carbamazepine at the 2010 CINP congress in Hong Kong when I was president of CINP, but Okuma could not come to receive his prize and died not long afterwards. The mood-stabilizing properties of valproate were discovered by Hinderk Emrich in Germany based on reports in French and German that valproic acid and its amide derivative valnoctamine were sedative. Emrich followed up his early case studies with small controlled trials, all published in German. It may be of significance that lithium valproate and carbamazepine were all discovered by non-Americans who were willing and able to do clinical exploration in patients based on little evidence except their personal and perhaps idiosyncratic theories. All three of them are heroes in my eyes: Okuma and Emrich as well as Cade. Carbamazepine was taken up by Robert Post at the National Institutes of Health (NIH) within a theoretical framework that epilepsy and bipolar disorder may represent aspects of the same kind of recurrent abnormal brain neuronal discharge as bipolar disorder. Post's studies are heuristic and interesting. However, few of his studies generated definitive efficacy data for carbamazepine in bipolar disorder. No pharmaceutical company was interested in a "use patent" for carbamazepine in bipolar disorder since it had already come "off-patent" for epilepsy. Therefore, the quality of research on carbamazepine's efficacy for bipolar disorder remains limited to this day. More recently, a derivative of carbamazepine, oxcarbazepine, was patented, and a large-scale trial showed its efficacy in prophylaxis in bipolar disorder. While some clinical scientists insist that clinical trial data must relate to each specific chemical entity as requiring separate proof, my scientific epistemology is based more on the recent book by Judea Pearl, *The Book of Why: The New Science of Cause and Effect* [9], in which Pearl strongly and convincingly argues that statistics only make sense after scientific hypotheses include biological causation theory; thus, I think the success of oxcarbazepine supports the efficacy of carbamazepine in bipolar disorder. By contrast with carbamazepine, the American FDA was willing to give a special-use patent to the valproate derivative called "divalproex sodium". Some people chuckle at this decision while others find it cynical and disturbing. Divalproex is a simple salt of valproic acid and dissolves in the stomach into valproate and is no more deserving of a special patent than lithium sulfate would be deserving of a separate set of studies comparing it to lithium carbonate. However, the FDA decision had the very positive effect of motivating the pharmaceutical company Abott to fund critical studies of valproate ("divalproex sodium") vs. lithium vs. placebo in bipolar disorder in several phases of the disease. These classic studies were led by Charles Bowden (of blessed memory). They showed that valproate had some efficacy in some phases of bipolar disorder and lithium did also, compared to placebo, but lithium was certainly not superior to valproate.

The BALANCE study is often seen as showing superiority of lithium to valproate [10]. The numbers were 31% episode-free for valproate vs. 41% for lithium: a barely statistically significant effect and not a decisive difference in my eyes. Not much specificity here. For a particular patient 41% vs. 31% is not as decisive as the side effect profile. BALANCE did not find clinical predictors. In his FDA pivotal 12-month study of divalproex and lithium, Bowden found superiority of valproate over lithium, which was not different from placebo in most measures. These are not to the point in any case because none of these facts replicate in the Kishi et al. [11] meta-analysis and in any case the differences between valproate and lithium in the aforementioned studies are small.

The model that I support is in the figure. We have here overlapping treatments, none of which are specific overall to the illness we call bipolar disorder. Sometimes a patient responds to one and sometimes to another.

5. Is Lithium Superior to Second-Generation Antipsychotics as a Mood Stabilizer?

When the second-generation antipsychotics became available, it was logical and obvious to try them in mania, since first-generation antipsychotics had been universally known to be highly effective in mania. Many studies had shown that the first-generation antipsychotics reduced activity and aggressiveness more rapidly than lithium in mania but others indicated that lithium in mania had greater patient acceptability because the patient felt that his mood improved more naturally and more gradually compared to the rapid straightjacket effect of first-generation antipsychotic treatment. Second-generation antipsychotics beginning with olanzapine seemed to reduce acute mania more like lithium clinically: there were few extra pyramidal side effects (EPS) and no sense of akinesia or thought emptying. There was more sedation than lithium, which proved a significant clinical advantage in acute mania. Few clinicians were prepared, however, for the fact that olanzapine and other second-generation antipsychotics turned out to be prophylactic in both phases of bipolar disorder [12]. The pharmaceutical industry, while its motives may be by definition financial, should be commended for not adhering to the boundaries of the *Diagnostic and Statistical Manual of Mental Disorders* (DSM) and for aggressively pursuing a program of extending by large clinical trials the indications of olanzapine and other second-generation antipsychotics into the realm of affective disorders. Olanzapine was found to be equally effective to lithium in the prophylaxis of mania and depression [13]. Even more surprisingly, olanzapine and particularly quietiapine, (Seroquel) were found to be equally effective to SSRIs in the treatment of acute bipolar depression as well as prophylaxis of bipolar depression. These new second-generation antipsychotics have numerous biochemical properties, many of which differ between the different second-generation antipsychotics. Most block the 5HT2 receptor as well as the D4 receptor but others such as amisulpride do not. Recent partial agonists/antagonists such as aripiprazole have little activity at the 5HT2 receptor but seem to be highly effective in the prophylaxis of bipolar disorder both in manic and depressive phase. It has not yet been determined whether second-generation antipsychotics that are D-2/5HT2 blockers are more efficacious in a different subgroup of bipolar patients than the partial agonists such as aripiprazole or even the D3 dominant partial agonists such as cariprazine or brexpiprazole. The DSM-5 emphasis on bipolar disorder as an entity encourages treatment trials with entry criteria of bipolar disorder and endpoints of results of the rating scale, but subgroup analysis becomes post hoc and statistically suspect. The DSM-5 concept of bipolar disorder has become so dominant that clinical intuition about possible subtypes that could then be tested with specific drugs has been effectively suppressed. It would be baffling if these very varied biochemical receptor acting drugs all worked in the same way on the heterogenous clinical syndrome of bipolar disorder (Figure 1).

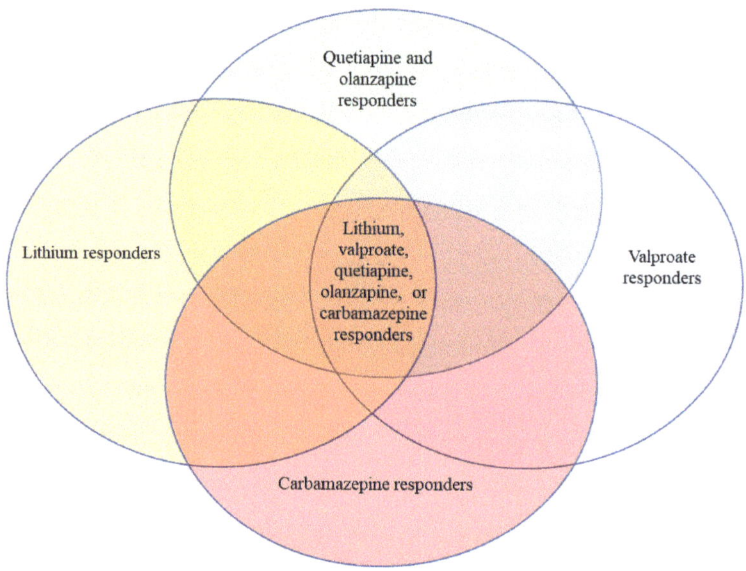

Figure 1. The above figure illustrates roughly the fact that some people respond to lithium, another group responds to valproate, another group responds to carbamazepine, and another group responds to some dopamine blockers; many patients with bipolar disorder respond to any of these four classic mood stabilizer groups, and some respond to only one or two. There is good clinical data to call all of these treatments mood stabilizers, but each is distinct chemically and biochemically. Printed with modifications and permission from Wolters Kluwer; Belmaker: "*Lithium*" in Kaplan and Sadock's *Comprehensive Textbook of Psychiatry*, edition 11, 2024 [14].

6. Were First-Generation D-2 Blockers Also Mood Stabilizers?

Some current clinicians are unaware that first-generation antipsychotics were probably mood stabilizers. See Bond et al., "Depot antipsychotic medications in bipolar disorder: a review of the literature" [15] or Littlejohn and Cookson: "Depot antipsychotics in the prophylaxis of bipolar affective disorder" [16]. Authors who have worked outside of the US and Europe know that antipsychotics are used by huge populations all around the world effectively for the treatment of bipolar disorder [17]. I have in my files a piece of correspondence about this issue, which I wrote 30 years ago to many leaders in the bipolar field. The responses that I have in my files, which have also been preserved in the CINP/American College of Neuropsychopharmacology (ACNP) historical archives of the history of psychopharmacology by Tom Ban, showed that no leader in the bipolar field in that time denied the efficacy of first-generation antipsychotics in bipolar disorder, including prophylaxis. They merely answered that the risk of tardive dyskinesia was too high to recommend (then atypical) antipsychotics as opposed to lithium. Moreover, historians of psychopharmacology have clearly pointed out the powerful influence of the intellectual needs of psychopharmacologists themselves in defining the nature of mood stabilizer vs. antipsychotic vs. antidepressant. As an emerging field, we wanted to find clinical/biochemical specificity. This historical pressure could be the subject of a whole article but I urge the reviewer to think about it. I cannot think of one drug that has dopamine receptor blocking properties, from clozapine to olanzapine to aripiprazole to cariprazine to lurasidone, that has not been shown to be an effective treatment for one or both poles of bipolar disorder without exacerbation of the other pole—the definition of a mood stabilizer, as I have read.

For some, the issue of tardive dyskinesia ends first-generation antipsychotics. This statement bothers me greatly in 2025. The second-generation antipsychotics have led to an epidemic of diabetes and obesity in bipolar patients and other patients that will be on our conscience for decades. The fact that first-generation antipsychotics cause tardive dyskinesia must be weighed against the terrible clinical side effects of second-generation antipsychotics in each patient specifically. I have written about it in my book *Psychopharmacology Reconsidered: A Concise Text Exploring the Limits of Diagnosis and Treatment* [18].

7. Do We Have a Mechanism of Action of Lithium Worth Teaching to Medical Students or Residents?

Thirty-five years ago, I co-authored an editorial [19] attempting to adjudicate between the two leading theories of that era: adenylate cyclase inhibition via G-proteins, and inositol monophosphatase (IMPase) inhibition. Both theories seemed promising based on the biochemical evidence available then. Neither has yielded the therapeutic insights or new treatments that we hoped for. This pattern—of initially exciting findings that fail to translate into clinical advances—has repeated itself across multiple generations of lithium research.

7.1. The Generational Pattern of Lithium Research

As I discussed in detail in a 2004 New England Journal of Medicine article [20], lithium research has exhibited a remarkable pattern: in each era of psychopharmacology, lithium has been found to affect whatever biochemical system was at the frontier of investigation. When monoamine metabolites dominated research in the 1960s and 70s, lithium was found to affect norepinephrine metabolism. When receptor studies emerged in the 1970s and 80s, lithium prevented dopamine receptor supersensitivity. When second messenger systems became central in the 1980s and 90s, lithium inhibited key enzymes in multiple messenger pathways. When gene expression studies dominated in the late 1990s and 2000s, lithium affected transcription factors. Most recently, when neurogenesis and neuroprotection became prominent, lithium was found to affect Bcl-2 and cell survival pathways.

This generational shifting might reflect lithium's genuine complexity as a simple ion affecting fundamental cellular processes. Alternatively, it might reveal more about the sociology of scientific research than about lithium's true mechanism of action. Each generation of researchers has found what they were equipped and incentivized to look for.

7.2. Key Biochemical Findings: A Critical Summary

Rather than exhaustively reviewing the biochemical literature—which I have performed comprehensively elsewhere [21]—I will critically evaluate the major mechanistic theories that have shaped lithium research and teaching.

The Cyclic AMP Hypothesis: Our laboratory demonstrated in the 1980s that lithium at therapeutic doses completely blocked the cyclic AMP response to epinephrine in euthymic bipolar patients [21], replicating in humans what had been found in animal systems. We developed receptor binding assays showing that lithium could inhibit G-protein activation in rat brain tissue [22]. However, this effect proved highly dependent on magnesium concentrations in the experimental medium, and its relevance to human therapeutic conditions remains unclear. More critically, if cyclic AMP inhibition were lithium's primary mechanism, we should have been able to develop clinically useful compounds that more specifically target this pathway. We have not.

The Inositol Depletion Hypothesis: Perhaps the most influential theory proposed that lithium's inhibition of inositol monophosphatase at therapeutic concentrations (Ki = 0.8 mM) depletes cellular inositol, particularly in overactive neurons, thereby providing specificity to lithium's action [23–25]. This elegant "inositol depletion" theory dominated lithium research for two decades. Our laboratory extensively studied this mechanism, demonstrating that inositol administration could stereospecifically reverse certain behavioral effects of lithium in rodents [26]. We developed knockout mouse models where inositol monophosphatase-1 knockout mice showed lithium-like behaviors [27].

However, this theory has significant problems. Most critically, when ebselen—a compound that inhibits inositol monophosphatase and lowers brain inositol—became available for testing, it failed to replicate lithium's effects in key behavioral models [28]. This negative finding, published by our laboratory in 2020, undermines decades of research built on the assumption that inositol depletion is the key to lithium's clinical effects. Additionally, inositol did not function as an "antidote" for lithium toxicity as the theory would predict [29], and several studies failed to demonstrate that lithium consistently reduces brain PIP2 concentrations in primates.

The GSK-3 Hypothesis: Glycogen synthase kinase-3 (GSK-3) emerged as an attractive target when lithium was found to directly inhibit this ubiquitous kinase at concentrations near the therapeutic range (2.1 ± 0.6 mM) (41). GSK-3 plays critical roles in multiple cellular processes including metabolism, gene expression, and apoptosis [30,31]. The finding that valproic acid also inhibits GSK-3 [32] initially strengthened the hypothesis by suggesting a common mechanism for mood stabilizers. Moreover, lithium's effects on development and Wnt signaling could be explained through GSK-3 inhibition [33,34].

However, the GSK-3 hypothesis faces the same translational problem as earlier theories: the inhibition occurs at the top of lithium's therapeutic range or higher [35], raising questions about whether this is a mechanism of therapeutic action or toxicity. Despite considerable investment, synthetic GSK-3 inhibitors have not been successfully developed as brain-penetrable, brain-selective compounds suitable for clinical testing as lithium alternatives. The failure to translate this biochemical finding into therapeutic application after more than two decades suggests fundamental limitations in our understanding.

Other Proposed Mechanisms: Lithium's effects on serotonin autoreceptors [36–38], on neuroprotective proteins like Bcl-2 [39–43], on PAP phosphatase [44–48], and on glutamate systems [49–53] have all generated research programs and publications. Each finding is replicable within specific experimental conditions. None has led to new treatments or to a consensus understanding of lithium's clinical action.

7.3. The Alzheimer's Controversy: A Cautionary Tale

The recent enthusiasm for lithium in Alzheimer's disease prevention illustrates the dangers of over-interpreting biochemical findings. Based on sophisticated animal studies showing GSK-3β effects on tau phosphorylation, researchers proposed that lithium might prevent amyloid deposition and delay Alzheimer's disease [54,55]. Press releases and media coverage were extensive. Many Alzheimer's patients and their families were encouraged to begin lithium treatment.

However, the lithium concentrations used in these studies were often 1000 times lower than any concentration ever shown to inhibit GSK-3 in biochemical assays—a fact that should have induced immediate skepticism. Claiming that a microdose (1/1000th of the usual therapeutic dose) of lithium can prevent Alzheimer's is as scientifically defensible as claiming that 1/1000th of the usual LSD dose prevents depression. The comparison to microdosing psychedelics is intentional: both claims reflect wishful thinking rather than rigorous science.

Well-controlled clinical trials of lithium in Alzheimer's disease have been predominantly negative. Yet, this failed hypothesis has been recycled to support broader claims about lithium as a "neuroprotective agent," creating an undeserved halo effect that obscures rather than illuminates lithium's actual clinical utility. It is especially troublesome that these researchers touted lithium orotate as a special compound. Li in medicine is a salt that dissolves in the stomach into Li^+ ions. Only Li is measured in blood. Orotate is metabolized.

7.4. Modern Approaches: Gene Expression and Stem Cell Studies

Recent research has taken hypothesis-free approaches using large datasets. Akkouh and colleagues [56] studied lithium's effects on thousands of genes simultaneously using gene chips, finding that lithium increases some mRNA levels while decreasing others. This is analogous to observing that lithium affects an orchestra's performance: you can see that some wind instruments become faster, some percussion becomes slower, and some violins change pitch—but this does not tell you what effect lithium is having on the conductor himself. To develop a new lithium-like drug, we would need to identify the proximal mechanism that causes these hundreds of simultaneous gene expression changes [57].

Stern and colleagues have pioneered the use of stem cells differentiated into neurons from skin biopsies of bipolar patients versus controls, as well as immortalized lymphocytes from both groups [58]. This innovative approach allows study of human tissue without invasive procedures and addresses whether lithium has different effects in bipolar lithium-responsive patients than in controls or non-responders. Complex effects on cell membrane excitability have been found, though which effects are relevant or specific remains unclear.

7.5. Why No Mechanism After 70 Years?

Several explanations warrant consideration:

Lithium may have no single proximal mechanism. As a simple ion affecting fundamental cellular processes involving sodium, potassium, and magnesium, lithium might exert its clinical effects through multiple, redundant pathways. This "many mechanisms" hypothesis contradicts Occam's razor but may reflect biological reality.

Bipolar disorder itself may be heterogeneous. If bipolar disorder as currently diagnosed represents multiple distinct pathophysiologies, then searching for a unifying mechanism of lithium action is misguided. Lithium may work through different mechanisms in different patient subgroups—an explanation that fits clinical observations better than the "one disease, one mechanism" model.

The therapeutic concentration is the problem. Unlike drugs that bind to specific receptors at nanomolar concentrations, lithium works at millimolar concentrations where it affects numerous systems simultaneously. This lack of specificity at the molecular level may make it impossible to identify a primary mechanism distinct from secondary effects.

Our experimental systems may be inadequate. Most biochemical studies use artificial in vitro systems or simplified animal models that may not capture the relevant clinical biology. The repeated failure to translate promising animal findings into human therapeutics suggests fundamental limitations in our research paradigms.

7.6. What Should We Teach Medical Students?

Given this state of affairs, what is worth teaching to medical students and residents about lithium's mechanism of action?

What we should NOT teach: We should not present any single mechanism as "the answer" to how lithium works. Teaching that lithium works primarily through inositol depletion, or GSK-3 inhibition, or any other single pathway, misrepresents the current state of knowledge and sets students up for confusion when they encounter contradictory information later.

What we SHOULD teach:

1. Honesty about uncertainty: Lithium's mechanism of action is unknown despite seven decades of intensive research. This is not a knowledge gap that will be filled by one more study—it represents a fundamental challenge in psychiatric neuroscience.
2. The pattern of research: Lithium has been found to affect virtually every biochemical system investigated at multiple levels, from ion channels to gene expression. This may reflect genuine biological complexity rather than inadequate research.
3. The contrast with other psychotropics: Unlike antipsychotics (dopamine D-2 blockade) or antidepressants (monoamine reuptake inhibition), we lack even a simplified textbook-level explanation of lithium's action that bears a defensible relationship to truth.
4. Clinical implications: The lack of a known mechanism means we have no laboratory test to predict lithium response, no rational approach to designing improved lithium-like compounds, and no way to identify which biochemical abnormality in bipolar disorder lithium is correcting.
5. The lesson for psychiatry: Lithium's story teaches humility. Effective treatments can exist without mechanistic understanding. Clinical efficacy and biochemical mechanism are separate questions. The assumption that understanding mechanism is "the key to understanding" a psychiatric disorder may itself be flawed.

7.7. Implications for the "Gold Standard" Debate

The absence of a known mechanism of action has important implications for lithium's status in treatment guidelines. Those who argue that lithium is the "gold standard" for bipolar disorder sometimes invoke its long research history and the extensive investigation of its biochemical effects as supporting evidence for its superiority. However, as this section demonstrates, 70 years of biochemical research has not revealed a mechanism of action, has not led to improved lithium-like compounds, and has not provided methods to predict which patients will respond.

The mystique surrounding lithium—as a simple ion that should be easy to understand, as a naturally occurring element with hypothesized specificity for bipolar disorder—has arguably hindered rather than helped clinical practice. It has encouraged an ideological attachment to lithium that sometimes overrides individualized clinical judgment about which treatment is best for a particular patient.

Basic scientists studying lithium biochemistry often begin their papers by declaring lithium "the gold standard for bipolar disorder" as justification for their research. But this large body of biochemical research, much of which has not been replicated and none of which has led to a proven mechanism of action, cannot be taken as proof of lithium's clinical superiority over other mood stabilizers. The lack of pharmaceutical company support for lithium research means that comparative studies with adequate statistical power have been limited. We simply do not know whether lithium's mood-stabilizing effects are more potent, more durable, or more specific than those of valproate, carbamazepine, or second-generation antipsychotics.

Summary: After 70 years of research, we must acknowledge that lithium is a clinically useful compound whose mechanism of action remains unknown. This is not a temporary gap in knowledge but may reflect fundamental complexity in both lithium's biology and bipolar disorder's pathophysiology. Rather than continuing to invest in mechanistic studies that fail to translate into clinical advances, we might better serve patients by focusing on comparative effectiveness research, prediction of individual treatment response, and development of truly novel therapeutic approaches based on emerging understanding of mood disorder neurobiology.

8. Does Response to Lithium Define Bipolar Disorder and Is Bipolar Disorder Always an Indication for Use of Lithium?

We do not know which cases of bipolar disorder will respond to lithium. Martin Alda and Paul Grof are two talented and wonderful researchers whom I respect greatly. I have gone over their papers for the umpteenth time and they find correlations that explain a tiny percentage of the variation in lithium response that can be explained by the clinical presentation or genetics. For instance, "Genetic variance with response to lithium treatment in bipolar disorder: a genome-wide association study" in *The Lancet* [59] studied 2563 patients collected by 22 participating sites from the International Consortium on Lithium Genetics and found a single locus associated with response to lithium monotherapy. Given that all data on lithium therapy in bipolar disorder finds no single Mendelian pattern, it is highly unlikely that this single locus explains a significant part of the variance [60]. Another paper by Paul Grof entitled "Sixty Years of Lithium Responders" [61] states in the abstract, "Despite promising reports, particularly from molecular genetics, we are still waiting for a biological elucidation of the stabilizing effect of lithium. The most useful predictor of lithium stabilization to date has been the patients' clinical profile based on a comprehensive clinical assessment; complete remission and other characteristics of episode clinical course, low psychiatric comorbidity...". This seems like the definition of a patient who would respond to other treatments as well to lithium: "the good prognosis patient". The paper presents no evidence that these phenomena predict lithium responders rather than just "responders". Grof goes on to say that the clinical picture of the lithium responders approximates the classical Kraepelinian description of a manic-depressive patient. NO, NO, NO! Kraepelin's definition of manic-depressive illness included patients with highly psychotic manias and highly psychotic depressions. These patients were clearly excluded from Grof's definition of good lithium responders but were included in the NIMH groups by Goodwin and others, showing that lithium response is not dependent on the presence or absence of severe psychosis in the mania and also many papers showing the excellent response of many schizoaffective patients to lithium [62]. Many clinicians seem to confuse two different concepts: Statement #1: Lithium works better in classical bipolar 1 patients than in other forms of psychiatric disorder. Statement #2: Lithium works better than valproate, carbamazepine, olanzapine, or quetiapine in classical bipolar 1 patients.

Statement #1 has some evidence from the Alda and Grof work, but contradicts much work of Goodwin at NIMH, such as Carlson and Goodwin's "The Stages of Mania" 1973 [62], where the presence of psychosis in mania does not predict lithium response. Statement #2 has not really been tested to my knowledge but it should be realized that classical bipolar 1 according to Grof and Alda is not Kraepelinian manic-depressive disorder nor is it a genetic subgroup of bipolar disorder. It is merely mild bipolar disorder with many good prognostic features. Perhaps statement #1 would be equally true of carbamazepine, valproate, and olanzapine in the current climate where bipolar 2 disorder is more frequent than bipolar 1 disorder and many borderline patients are diagnosed as bipolar. Without comparing the lithium efficacy to that of valproate and dopamine blockers, this whole area of research has gone down a rabbit hole.

Some say that Kishi et al. [11,63] contradict other meta-analyses demonstrating lithium's superiority over placebo. These scholars are incorrect. They are answering the wrong question. No one has ever claimed in recent decades that lithium is not better than placebo. The point studied by Kishi et al. [11,63] is whether it is better than compounds with completely different mechanisms of action such as the antiepileptics and the dopamine receptor blockers.

Some say that it is critical to compare data quality rather than efficacy. There is some sense in which this idea is correct: any compound that has drug company backing will most likely have large short-term studies. Lithium data, however, has much smaller sample sizes and has rarely been studied with modern multicentered designs. I think that this is comparing apples and oranges and I stand behind my statement, "Lithium is only one of many effective treatments for bipolar disorder". We do not have a football game-style design as to whether I am for or against lithium. I am not against lithium and agree that the mechanism of many, if not all, psychiatric treatments is unknown. Lithium is not uniquely unfathomable, but it is not a mystically heuristic holy grail either. The point is that lithium was claimed over these last 75 years to be such a simple ion that we would soon understand it. This has not turned out to be the case and it is time to face up to that fact. I do not agree that such dramatic effects as lithium produces are rarely seen in treatments with other mood stabilizers. I have seen bipolar patients respond wonderfully over many years to carbamazepine, to valproate, to olanzapine, and yes, to 1 mg of haloperidol.

9. Lithium and Suicidality

A large number of epidemiological studies support the concept that patients on lithium have fewer serious suicide attempts and fewer successful suicides. In and of itself, this finding bears only tenuous relationship to lithium's effects in bipolar disorder. True, bipolar disorder carries a high risk of suicide. But so do schizophrenia, unipolar disorder, and borderline personality disorder. Lithium's effect on suicidality may be an entirely independent effect similar to the hypothesized effect of ivermectin in COVID, although ivermectin was originally purposed as an antiparasitic drug. On the other hand, an epiphenomenon may be at work here. The only other drug reported in large epidemiological studies to have an anti-suicidal effect is clozapine. Clozapine has almost no biochemical similarities to lithium. The only similarity is that both clozapine and lithium require frequent blood monitoring. Perhaps the effects of lithium on suicidality are merely an epiphenomenon of the increased patient surveillance required by the lithium blood testing. Many such associations exist in the medical literature and I find it an unconvincing fact when it is used in support of lithium's uniqueness.

10. The Dangers of the Intellectual Cult Around Lithium and the Importance of an Empirical Patient-Centered Therapeutic Armamentarium

A new book about critical psychiatry by Awais Aftab (*Conversations in Critical Psychiatry, Oxford University Press*, ref. [64]) discussed the importance of science's ability to self-criticize and the particular importance of this for psychiatry. Admirably, he has emphasized that critics also need to be self-critical. There is no question that an "antipsychiatry movement" exists that denies the existence of mental illness, specific diagnoses, or the possibility that chemical treatments would benefit individuals with mental disorders. This antipsychiatry movement is not self-critical and may not be possible to engage in a beneficial fashion. However, not all critiques of psychiatry and psychopharmacology are tainted with ideological antipsychiatry. The writings of Montcrief [65,66] have pointed out that antidepressants may be blunting all emotion rather than specifically lifting abnormal depressed mood. This hypothesis is far from proven but worthy of serious consideration in psychopharmacology. Robin Murray [67] and equally serious psychopharmacological scholars also question whether dopamine-blocking drugs are truly antipsychotic or whether their therapeutic benefit derives from restricting human imagination and creative thinking and thereby including subjective impairments that are a serious cost for the patient under treatment. The possibility that antidepressant and antipsychotic treatments are working by helping some patients but inducing a cost in others leads to a therapeutic calculus that is very different than an approach based on the idea that psychopharmacology treats specific diseases with specific treatments. For some critics of psychopharmacology, lithium has remained as an exception, a rare example of a treatment which has been assumed because of its biochemical specificity to be truly related to the pathophysiology of bipolar disorder. I do not think that this has been proved. The need for studies of the effects of lithium on the personality of the bipolar patient who benefits from it is great. Even in those patients where the benefit greatly outweighs the psychological or medical costs and side effects, there is no reason to discount in advance the complaints of many patients that lithium restricts their range of affective expression or creativity. All medical and psychiatric treatment is a balance between hoped-for benefits and appropriately considered side effects of any external intervention. Lithium should not be seen as an exception because the rule of this cost/benefit analysis is the very basis of medicine and psychiatry.

I still use lithium in the clinic and find it very beneficial in many patients with bipolar disorder, some with recurrent unipolar disorder and many with schizoaffective disorders; atypical antipsychotics are often but not always equally useful and are very convenient [68]. Given the long period of lithium usage in many of my patients, kidney effects including progression to renal dialysis and transplant have become a very real issue [69]. Basic research on the mechanism of lithium action has not led to application of Occam's razor, but clinical usage surely justifies the adage that there is no free lunch.

This review is a response to the exuberance of the Mahli and Bauer editorial [4]. Such exuberance is in such contrast to the practice seen by residents in psychiatry everywhere that it leads to confusion and undermines the legitimacy of academic leadership. I agree that a drug's specificity is not directly related to its efficacy. In the case of lithium, the history of psychopharmacology connected the two and this is the reason that I address the two together. Lithium's use as a gold standard has generated thousands of articles using lithium in the test tube presenting results as possible mechanisms of bipolar disorder. My claim is that such biochemical studies should use lithium, an antiepileptic, and a dopamine blocker in each study if any inferences about the underlying illness are to be made. I think the history of this field has indeed been damaged by the idea that lithium is "the gold standard".

Lithium can play a treatment role in a variety of conditions. Some praise the clinical experience of lithium in the context of bipolar disorder over the "long term". I do not know of any long-term controlled studies of lithium vs. placebo. All long-term studies are of selected patients followed based on various clinical and genetic phenotypes. I do not agree that lithium has the largest database. In fact, lithium just 25 years ago was a subject of controversy as to whether it worked at all until lithium was used as a control treatment in several large pharmaceutical company-sponsored trials of other compounds. Most of the lithium database in the long term uses small sample sizes in highly selected patient groups.

Long-term good lithium response could equally well be a predictor of natural course of illness and would be equally valid with any other medication if tried [70]. The original, most gripping description of bipolar disorder, *A Mind that Found Itself* by Clifford Beers in 1908 [71], was of a patient who had five years of continuous hospitalization with severe bipolar mood alterations that ended with complete recovery and a lifetime of productive social activities, long before the introduction of lithium.

Funding: This research received no external funding.

Institutional Review Board Statement: Not applicable.

Informed Consent Statement: Not applicable.

Data Availability Statement: No new data were created or analyzed in this study. Data sharing is not applicable to this article.

Acknowledgments: During the preparation of this manuscript/study, the author used AI Claude Opus 4.1 for the purposes of restyled and rephrased Section 7. The authors have reviewed and edited the output and take full responsibility for the content of this publication.

Conflicts of Interest: The author declares no conflicts of interest.

References

1. Shorter, E. *The Rise and Fall of the Age of Psychopharmacology*; Oxford University Press: Oxford, UK, 2021.
2. Ban, T. *Lithium in Psychiatry in Historical Perspective*; INHN Publishers: Córdoba, Argentina, 2021.
3. Brown, W. *Lithium: A Doctor, a Drug, and a Breakthrough*, 1st ed.; Liveright: New York, NY, USA, 2019.
4. Malhi, G.S.; Bauer, M. Lithium first: Not merely first line. *Bipolar Disord.* **2023**, *25*, 7–8. [CrossRef]
5. Chan, J.K.; Solmi, M.; Correll, C.U.; Wong, C.S.; Lo, H.K.; Lai, F.T.; Chang, W.C. Lithium for Bipolar Disorder and Risk of Thyroid Dysfunction and Chronic Kidney Disease. *JAMA Netw. Open* **2025**, *8*, e2458608. [CrossRef]
6. Chan, J.K.; Solmi, M.; Correll, C.U.; Wong, C.S.; Lo, H.K.; Lai, F.T.; Chang, W.C. Predicting 10-year risk of chronic kidney disease in lithium-treated patients with bipolar disorder: A risk model development and internal cross-validation study. *Eur. Neuropsychopharmacol.* **2025**, *95*, 24–30. [CrossRef] [PubMed]
7. Greil, W.; de Bardeci, M.; Nievergelt, N.; Erfurth, A.; Hasler, G.; Bridler, R.; Toto, S.; Grohmann, R.; Seifert, J.; Schoretsanitis, G. Twenty-four years of prescription patterns in bipolar disorder inpatients with vs without lithium: A pharmacoepidemiological analysis of 8707 cases in German-speaking countries. *Int. J. Bipolar Disord.* **2025**, *13*, 3. [CrossRef] [PubMed]
8. Okuma, T.; Kishimoto, A.; Inoue, K.; Matsumoto, H.; Ogura, A. Anti-manic and prophylactic effects of carbamazepine (Tegretol) on manic depressive psychosis. A preliminary report. *Folia Psychiatr. Neurol. Jpn.* **1973**, *27*, 283–297. [CrossRef] [PubMed]
9. Pearl, J.; Mackenzie, D. *The Book of Why: The New Science of Cause and Effect*; Basic Books: New York, NY, USA, 2018.
10. Geddes, J.R.; Goodwin, G.M.; Rendell, J.; Azorin, J.M.; Cipriani, A.; Ostacher, M.J.; Alder, N.; Morriss, R.; Juszczak, E. Lithium plus valproate combination therapy versus monotherapy for relapse prevention in bipolar I disorder (BALANCE): A randomised open-label trial. *Lancet* **2010**, *375*, 385–395. [CrossRef]
11. Kishi, T.; Ikuta, T.; Matsuda, Y.; Sakuma, K.; Okuya, M.; Nomura, I.; Hatano, M.; Iwata, N. Pharmacological treatment for bipolar mania: A systematic review and network meta-analysis of double-blind randomized controlled trials. *Mol. Psychiatry* **2022**, *27*, 1136–1144. [CrossRef]
12. Cleare, A.J.; Kerr-Gaffney, J.; Goldsmith, K.; Zenasni, Z.; Yaziji, N.; Jin, H.; Colasanti, A.; Geddes, J.R.; Kessler, D.; McAllister-Williams, R.H.; et al. Clinical and cost-effectiveness of lithium versus quetiapine augmentation for treatment-resistant depression: A pragmatic, open-label, parallel-group, randomised controlled superiority trial in the UK. *Lancet Psychiatry* **2025**, *12*, 276–288. [CrossRef]

13. Ermis, C.; Taipale, H.; Tanskanen, A.; Vieta, E.; Correll, C.U.; Mittendorfer-Rutz, E.; Tiihonen, J. Real-world effectiveness of pharmacological maintenance treatment of bipolar depression: A within-subject analysis in a Swedish nationwide cohort. *Lancet Psychiatry* **2025**, *12*, 198–207. [CrossRef]
14. Belmaker, R.H. Lithium. In *Kaplan & Sadock's Comprehensive Textbook of Psychiatry*, 11th ed.; Boland, R., Verduin, M., Eds.; Wolters Kluwer: Philadelphia, PA, USA, 2024.
15. Bond, D.J.; Pratoomsri, W.; Yatham, L.N. Depot antipsychotic medications in bipolar disorder: A review of the literature. *Acta Psychiatr. Scand. Suppl.* **2007**, *434*, 3–16. [CrossRef]
16. Littlejohn, R.; Leslie, F.; Cookson, J. Depot antipsychotics in the prophylaxis of bipolar affective disorder. *Br. J. Psychiatry* **1994**, *165*, 827–829. [CrossRef]
17. Adiukwu, F.N.; Salihu, M.O.; Hayatudeen, N.; Kareem, Y.A.; Adegoke, B.O.; Charles-Ugwuagbo, I.C.; Uwah, E.E.; Adesina, I.O.; Ibrahim, A.M.; Fugu, M.A. Challenges of lithium prescription in bipolar disorders in Nigeria. *BMC Psychiatry* **2025**, *25*, 304. [CrossRef]
18. Belmaker, R.; Lichtenberg, P. *Psychopharmacology Reconsidered: A Concise Guide Exploring the Limits of Diagnosis and Treatment*; Springer: Cham, Switzerland, 2023.
19. Belmaker, R.H.; Kofman, O. Lithium research: State of the art. *Biol. Psychiatry* **1990**, *27*, 1279–1281. [CrossRef]
20. Belmaker, R.H. Bipolar disorder. *N. Engl. J. Med.* **2004**, *351*, 476–486. [CrossRef]
21. Ebstein, R.; Belmaker, R.; Grunhaus, L.; Rimon, R. Lithium inhibition of adrenaline-stimulated adenylate cyclase in humans. *Nature* **1976**, *259*, 411–413. [CrossRef]
22. Avissar, S.; Schreiber, G.; Danon, A.; Belmaker, R.H. Lithium inhibits adrenergic and cholinergic increases in GTP binding in rat cortex. *Nature* **1988**, *331*, 440–442. [CrossRef]
23. Belmaker, R.H.; Bersudsky, Y.; Agam, G.; Levine, J.; Kofman, O. How does lithium work on manic depression? Clinical and psychological correlates of the inositol theory. *Annu. Rev. Med.* **1996**, *47*, 47–56. [CrossRef] [PubMed]
24. Berridge, M.J. The Albert Lasker Medical Awards. Inositol trisphosphate, calcium, lithium, and cell signaling. *JAMA* **1989**, *262*, 1834–1841. [CrossRef] [PubMed]
25. Hallcher, L.M.; Sherman, W.R. The effects of lithium ion and other agents on the activity of myo-inositol-1-phosphatase from bovine brain. *J. Biol. Chem.* **1980**, *255*, 10896–10901. [CrossRef] [PubMed]
26. Kofman, O.; Belmaker, R.H. Ziskind-Somerfeld Research Award 1993. Biochemical, behavioral, and clinical studies of the role of inositol in lithium treatment and depression. *Biol. Psychiatry* **1993**, *34*, 839–852. [CrossRef]
27. Agam, G.; Bersudsky, Y.; Berry, G.T.; Moechars, D.; Lavi-Avnon, Y.; Belmaker, R.H. Knockout mice in understanding the mechanism of action of lithium. *Biochem. Soc. Trans.* **2009**, *37 Pt 5*, 1121–1125. [CrossRef] [PubMed]
28. Singh, N.; Serres, F.; Toker, L.; Sade, Y.; Blackburn, V.; Batra, A.S.; Saiardi, A.; Agam, G.; Belmaker, R.H.; Sharp, T.; et al. Effects of the putative lithium mimetic ebselen on pilocarpine-induced neural activity. *Eur. J. Pharmacol.* **2020**, *883*, 173377. [CrossRef] [PubMed]
29. Bersudsky, Y.; Vinnitsky, I.; Ghelber, D.; Kofman, O.; Kaplan, Z.; Belmaker, R.H. Mechanism of lithium lethality in rats. *J. Psychiatr. Res.* **1993**, *27*, 415–422. [CrossRef]
30. Cui, H.; Meng, Y.; Bulleit, R.F. Inhibition of glycogen synthase kinase 3beta activity regulates proliferation of cultured cerebellar granule cells. *Brain Res. Dev. Brain Res.* **1998**, *111*, 177–188. [CrossRef]
31. Salinas, P.C. Wnt factors in axonal remodelling and synaptogenesis. *Biochem. Soc. Symp.* **1999**, *65*, 101–109.
32. Chen, G.; Huang, L.D.; Jiang, Y.M.; Manji, H.K. The mood-stabilizing agent valproate inhibits the activity of glycogen synthase kinase-3. *J. Neurochem.* **1999**, *72*, 1327–1330. [CrossRef]
33. Dierick, H.; Bejsovec, A. Cellular mechanisms of wingless/Wnt signal transduction. *Curr. Top. Dev. Biol.* **1999**, *43*, 153–190.
34. Hedgepeth, C.M.; Conrad, L.J.; Zhang, J.; Huang, H.C.; Lee, V.M.; Klein, P.S. Activation of the Wnt signaling pathway: A molecular mechanism for lithium action. *Dev. Biol.* **1997**, *185*, 82–91. [CrossRef]
35. Stambolic, V.; Ruel, L.; Woodgett, J.R. Lithium inhibits glycogen synthase kinase-3 activity and mimics wingless signalling in intact cells. *Curr. Biol.* **1996**, *6*, 1664–1668. [CrossRef]
36. Haddjeri, N.; Szabo, S.T.; de Montigny, C.; Blier, P. Increased tonic activation of rat forebrain 5-HT_{1A} receptors by lithium addition to antidepressant treatments. *Neuropsychopharmacology* **2000**, *22*, 346–356. [CrossRef]
37. Massot, O.; Rousselle, J.C.; Fillion, M.P.; Januel, D.; Plantefol, M.; Fillion, G. 5-HT1B receptors: A novel target for lithium. Possible involvement in mood disorders. *Neuropsychopharmacology* **1999**, *21*, 530–541. [CrossRef]
38. Redrobe, J.P.; Bourin, M. Evidence of the activity of lithium on 5-HT1B receptors in the mouse forced swimming test: Comparison with carbamazepine and sodium valproate. *Psychopharmacology* **1999**, *141*, 370–377. [CrossRef]
39. Chen, D.F.; Schneider, G.E.; Martinou, J.C.; Tonegawa, S. Bcl-2 promotes regeneration of severed axons in mammalian CNS. *Nature* **1997**, *385*, 434–439. [CrossRef]
40. Chen, G.; Zeng, W.Z.; Yuan, P.X.; Huang, L.D.; Jiang, Y.M.; Zhao, Z.H.; Manji, H.K. The mood-stabilizing agents lithium and valproate robustly increase the levels of the neuroprotective protein bcl-2 in the CNS. *J. Neurochem.* **1999**, *72*, 879–882. [CrossRef]

41. Chen, R.W.; Chuang, D.M. Long term lithium treatment suppresses p53 and Bax expression but increases Bcl-2 expression. A prominent role in neuroprotection against excitotoxicity. *J. Biol. Chem.* **1999**, *274*, 6039–6042. [CrossRef]
42. Manji, H.K.; Moore, G.J.; Chen, G. Lithium at 50: Have the neuroprotective effects of this unique cation been overlooked? *Biol. Psychiatry* **1999**, *46*, 929–940. [CrossRef]
43. Manji, H.K.; Moore, G.J.; Chen, G. Lithium up-regulates the cytoprotective protein Bcl-2 in the CNS in vivo: A role for neurotrophic and neuroprotective effects in manic depressive illness. *J. Clin. Psychiatry* **2000**, *61* (Suppl. 9), 82–96. [PubMed]
44. Dichtl, B.; Stevens, A.; Tollervey, D. Lithium toxicity in yeast is due to the inhibition of RNA processing enzymes. *EMBO J.* **1997**, *16*, 7184–7195. [CrossRef] [PubMed]
45. Klassen, C.; Boles, J. The importance of 3′-phosphoadenosine 5′-phosphosulphate (PAPS) in the regulation of sulfation. *FASEB J.* **1997**, *11*, 404–417. [CrossRef]
46. López-Coronado, J.M.; Bellés, J.M.; Lesage, F.; Serrano, R.; Rodríguez, P.L. A novel mammalian lithium-sensitive enzyme with a dual enzymatic activity, 3′-phosphoadenosine 5′-phosphate phosphatase and inositol-polyphosphate 1-phosphatase. *J. Biol. Chem.* **1999**, *274*, 16034–16039. [CrossRef]
47. Spiegelberg, B.D.; Xiong, J.P.; Smith, J.J.; Gu, R.F.; York, J.D. Cloning and characterization of a mammalian lithium-sensitive bisphosphate 3′-nucleotidase inhibited by inositol 1,4-bisphosphate. *J. Biol. Chem.* **1999**, *274*, 13619–13628. [CrossRef]
48. Yenush, L.; Bellés, J.M.; López-Coronado, J.M.; Gil-Mascarell, R.; Serrano, R.; Rodríguez, P.L. A novel target of lithium therapy. *FEBS Lett.* **2000**, *467*, 321–325. [CrossRef]
49. Dixon, J.F.; Hokin, L.E. Lithium acutely inhibits and chronically up-regulates and stabilizes glutamate uptake by presynaptic nerve endings in mouse cerebral cortex. *Proc. Natl. Acad. Sci. USA* **1998**, *95*, 8363–8368. [CrossRef]
50. Dixon, J.F.; Lee, C.H.; Los, G.V.; Hokin, L.E. Lithium enhances accumulation of [3H]inositol radioactivity and mass of second messenger inositol 1,4,5-trisphosphate in monkey cerebral cortex slices. *J. Neurochem.* **1992**, *59*, 2332–2335. [CrossRef] [PubMed]
51. Dixon, J.F.; Los, G.V.; Hokin, L.E. Lithium stimulates glutamate "release" and inositol 1,4,5-trisphosphate accumulation via activation of the N-methyl-D-aspartate receptor in monkey and mouse cerebral cortex slices. *Proc. Natl. Acad. Sci. USA* **1994**, *91*, 8358–8362. [CrossRef] [PubMed]
52. Kennedy, E.D.; Challiss, R.A.; Nahorski, S.R. Lithium reduces the accumulation of inositol polyphosphate second messengers following cholinergic stimulation of cerebral cortex slices. *J. Neurochem.* **1989**, *53*, 1652–1655. [CrossRef] [PubMed]
53. Lee, C.H.; Dixon, J.F.; Reichman, M.; Moummi, C.; Los, G.; Hokin, L.E. Li+ increases accumulation of inositol 1,4,5-trisphosphate and inositol 1,3,4,5-tetrakisphosphate in cholinergically stimulated brain cortex slices in guinea pig, mouse and rat. The increases require inositol supplementation in mouse and rat but not in guinea pig. *Biochem. J.* **1992**, *282*, 377–385. [CrossRef]
54. Aron, L.; Ngian, Z.K.; Qiu, C.; Choi, J.; Liang, M.; Drake, D.M.; Hamplova, S.E.; Lacey, E.K.; Roche, P.; Yuan, M.; et al. Lithium deficiency and the onset of Alzheimer's disease. *Nature* **2025**, *645*, 712–721. [CrossRef]
55. De-Paula, V.J.R.; Radanovic, M.; Forlenza, O.V. Lithium and neuroprotection: A review of molecular targets and biological effects at subtherapeutic concentrations in preclinical models of Alzheimer's disease. *Int. J. Bipolar Disord.* **2025**, *13*, 16. [CrossRef]
56. Akkouh, I.A.; Skrede, S.; Holmgren, A.; Ersland, K.M.; Hansson, L.; Bahrami, S.; Andreassen, O.A.; Steen, V.M.; Djurovic, S.; Hughes, T. Exploring lithium's transcriptional mechanisms of action in bipolar disorder: A multi-step study. *Neuropsychopharmacology* **2020**, *45*, 947–955. [CrossRef]
57. Scholl, J.; Panchal, P.; Nelissen, N.; Atkinson, L.Z.; Kolling, N.; Saunders, K.E.; Geddes, J.; Rushworth, M.F.; Nobre, A.C.; Harrison, P.J.; et al. Neural signatures of risk-taking adaptations across health, bipolar disorder, and lithium treatment. *Mol. Psychiatry* **2025**, *30*, 2955–2965. [CrossRef] [PubMed]
58. Stern, S.; Santos, R.; Marchetto, M.C.; Mendes, A.P.; Rouleau, G.A.; Biesmans, S.; Wang, Q.W.; Yao, J.; Charnay, P.; Bang, A.G.; et al. Neurons derived from patients with bipolar disorder divide into intrinsically different sub-populations of neurons, predicting the patients' responsiveness to lithium. *Mol. Psychiatry* **2018**, *23*, 1453–1465. [CrossRef]
59. Hou, L.; Heilbronner, U.; Degenhardt, F.; Adli, M.; Akiyama, K.; Akula, N.; Ardau, R.; Arias, B.; Backlund, L.; Banzato, C.E.; et al. Genetic variants associated with response to lithium treatment in bipolar disorder: A genome-wide association study. *Lancet* **2016**, *387*, 1085–1093. [CrossRef] [PubMed]
60. Paribello, P.; Isayeva, U.; Pisanu, C.; Squassina, A.; Manchia, M. Pharmacogenomics and response to lithium in bipolar disorder. *Pharmacogenomics* **2024**, *25*, 689–706. [CrossRef] [PubMed]
61. Grof, P. Sixty years of lithium responders. *Neuropsychobiology* **2010**, *62*, 8–16. [CrossRef]
62. Carlson, G.A.; Goodwin, F.K. The stages of mania. A longitudinal analysis of the manic episode. *Arch. Gen. Psychiatry* **1973**, *28*, 221–228. [CrossRef]
63. Kishi, T.; Ikuta, T.; Matsuda, Y.; Sakuma, K.; Okuya, M.; Mishima, K.; Iwata, N. Mood stabilizers and/or antipsychotics for bipolar disorder in the maintenance phase: A systematic review and network meta-analysis of randomized controlled trials. *Mol. Psychiatry* **2021**, *26*, 4146–4157. [CrossRef]
64. Aftab, A. *Conversations in Critical Psychiatry*; Oxford University Press: Oxford, UK, 2024.
65. Moncrieff, J. *The Myth of the Chemical Cure: A Critique of Psychiatric Drug Treatment*; Palgrave Macmillan: Basingstoke, UK, 2007.

66. Moncrieff, J. *The Bitterest Pills: The Troubling Story of Antipsychotic Drugs*; Palgrave Macmillan: Basingstoke, UK, 2013.
67. Murray, R.M. Mistakes I Have Made in My Research Career. *Schizophr. Bull.* **2017**, *43*, 253–256. [CrossRef]
68. Osher, Y.; Bersudsky, Y.; Belmaker, R.H. The new lithium clinic. *Neuropsychobiology* **2010**, *62*, 17–26. [CrossRef]
69. Azab, A.N.; Shnaider, A.; Osher, Y.; Wang, D.; Bersudsky, Y.; Belmaker, R.H. Lithium nephrotoxicity. *Int. J. Bipolar Disord.* **2015**, *3*, 28. [CrossRef]
70. Scott, K.; Khayachi, A.; Alda, M.; Nunes, A. Prediction of Treatment Outcome in Bipolar Disorder: When Can We Expect Clinical Relevance? *Biol. Psychiatry* **2025**, *98*, 285–292. [CrossRef] [PubMed]
71. Beers, C. *A Mind That Found Itself*, 1st ed.; Longmans, Green & Co.: New York, NY, USA, 1908.

pharmaceuticals

MDPI

Review

Could Lithium Be Preserved for the Stabilization of Bipolar Patients?

Paul Grof

Department of Psychiatry, Toronto University, Toronto, ON M5T 1R8, Canada; paulgrof75@gmail.com

Abstract

Lithium remains endorsed as first-line treatment for bipolar disorders across major clinical guidelines, yet robust evidence demonstrates its progressive decline in use in psychiatric practice across numerous countries. To justify this decline, concerns regarding lithium's efficacy, safety profile, and monitoring requirements are frequently cited. Yet these apprehensions largely stem from an insufficient understanding of lithium's clinical uses. In fact, when patients are selected for lithium stabilization according to a characteristic clinical profile and not just a bipolar verdict, lithium continues demonstrating good efficacy compared to all other psychiatric medications currently available. Moreover, after sufficient clinician and patient education regarding lithium stabilization principles, monitoring requirements stop being burdensome. Furthermore, among lithium-responsive patients, adverse effects are typically mild and clinically manageable, except for glomerular filtration rate decline, which tends to develop after decades of continuous administration. Thus, it might be possible to reverse this unfortunate decline in lithium's use by teaching clinicians to identify the patient profile responsive to lithium stabilization and by implementing educational programs regarding optimal lithium utilization for psychiatrists, patients, and their families. Furthermore, it is worth investigating intermittent lithium administration to mitigate renal complications.

Keywords: bipolar disorders; lithium; side-effects; monitoring

1. The Declining Use of Lithium

There was then and still is now no theoretical basis for lithium's use, no rationale that could then or can now be used to sell it. As a result, the use of lithium will almost certainly end when Schou dies. Another agent, probably of lesser efficacy, will displace it by virtue of a marketing strategy that depends on offering a "biological rationale"...

David Healy, MD, The Creation of Psychopharmacology, 2002.

Fortunately, David Healy's prophesy was not correct, but lithium's use has certainly been moving in this direction. What can we, clinicians and researchers, do to reverse this trend?

Since the 1960s, lithium treatment has protected countless lives of patients suffering from bipolar disorders [1] and normalized psychosocial functioning, and has generated substantial healthcare savings [2]. However, recent decades have witnessed a paradoxical situation: while clinical guidelines consistently describe lithium as first-line treatment for bipolar disorders, its utilization has progressively declined across multiple countries [3,4]. The evidence of this decline is robust and highlights significant obstacles to continued adoption [5,6]. There may be, however, other factors indirectly influencing the decreasing therapeutic use of lithium. These factors include increasing lithium's industrial use [7].

https://doi.org/10.3390/ph19040527

The reduction in lithium use has been particularly dramatic in the United States and pronounced throughout Western countries. However, lithium continues to be utilized in developing nations, presumably because it is inexpensive. For instance, recent data from the United States demonstrate that lithium use among patients with bipolar disorder decreased significantly from greater than 30% to below 15% over the past three decades. Although the European decline is less pronounced [3], both regions exhibit similar downward prescribing trends.

The main factors underlying this trend include concerns regarding lithium's adverse effects [8], unjustified but lingering doubts about its efficacy [9], the need for closer monitoring [10], and perceived management difficulties. Additional factors include aggressive marketing of second-generation antipsychotics, which demonstrate comparable efficacy in treating acute manic episodes, and limitations in lithium's commercial appeal [11]. Clinician preferences and attitudes toward lithium have been further influenced by patient and professional misinformation and beliefs [5].

These factors combined contribute to the decreasing trend in lithium utilization for treating bipolar disorder and related conditions. While lithium use has experienced temporary fluctuations previously following misguided publications between 1978 [12] and 2000 [13]—the current trend appears more sustained and ominous. If this trajectory continues without successful intervention, millions of patients with bipolar disorder and their families will suffer unnecessarily, with thousands never achieving stability and potentially losing their lives.

2. The Decline Likely to Worsen

Without intending to sound alarmist, there are justified reasons to anticipate that this decline in lithium use may accelerate. During the 1970s and 1980s, thousands of patients enthusiastically initiated lithium maintenance therapy. But after decades of continuous treatment, many of these patients begun exhibiting adverse renal effects [14].

Consequently, clinicians became increasingly aware of these complications, recognized the renal adversity in greater numbers of patients, and expressed concern regarding patient welfare, as well as potential medicolegal challenges. These developments have the potential to speed up lithium's disappearance from clinical practice, particularly in Western countries.

3. Misunderstood Efficacy

Over the years, the efficacy of lithium stabilization has been repeatedly questioned because of a misunderstanding. The underlying confusion stems from the radical shift in diagnosing bipolar disorders after the discovery of lithium's stabilizing value. Incontrovertible evidence of lithium's efficacy derived from double-blind trials was conducted by investigators who were trained to diagnose according to the Kraepelinian tradition [15]. Those clinicians identified manic-depressive illness according to the patient's clinical profile, with particular attention to the fully episodic course and supportive findings in family history.

But later, the diagnostic fashion changed. Diagnosing bipolar disorder exclusively according to the DSM symptom-based criteria [16] (Diagnostic and Statistical Manual of Mental Disorders) gradually prevailed and very significantly expanded diagnostic boundaries. It is relevant to explain briefly how this shift unfolded and changed lithium's use [17,18].

When Schou and Baastrup first reported lithium's remarkable stabilizing effects [19,20], no other medication could influence manic-depressive recurrences. Once lithium stabilization was approved by regulatory agencies, suddenly, regained mood stability was observed in thousands of manic-depressive patients who had previously suffered for decades, spend-

https://doi.org/10.3390/ph19040527

ing considerable time hospitalized. With lithium maintenance, these patients suddenly returned to their families and work, maintaining long-term stability—an outcome that then seemed miraculous.

Naturally, this extraordinary observation generated much hope, as well as extensive experimentation, in patients with difficult-to-treat psychiatric disorders. With much optimism, many such illnesses were tried on lithium. Lithium was subsequently evaluated in patient cohorts with 15 different diagnoses [21]. Most did not respond to lithium. However, complete stabilization was achieved in classical manic-depressive illness, with some benefits detected also in schizoaffective and schizophreniform disorders [22–24] and cycloid psychoses [25]—three conditions sharing mood abnormalities in their clinical presentation. The conclusion drawn was that these three conditions represent misdiagnosed bipolar disorders, with any lithium benefit wrongly equated to bipolar disorder.

In parallel with the change in diagnostic fashion, lithium's apparent efficacy gradually diminished and eventually seemingly vanished [13,17]. This oversimplified diagnostic reasoning (diagnosis by DSM symptoms as opposed to the patient's clinical profile) unfortunately overlooked that lithium offers psychiatric patients a variety of ten benefits that are qualitatively different from the stability achieved in the classical type of manic-depressive patients [26]. For example, in the episodically recurring bipolar type, lithium prevents both manias and depressions successfully; discontinuation produces no rebound phenomenon [27,28], the lithium response is fully reproducible after restarting lithium, and the anti-suicide effect is well documented [29–31].

Patients with other bipolar spectrum subtypes benefit differently: lithium only helps with manias but not depressions, discontinuation often precipitates rapid and exaggerated rebound, and the gain from lithium becomes poorly reproducible after discontinuation [32]. The benefits more closely resemble those of neuroleptics.

These qualitative differences were unfortunately ignored, and the diagnostic fashion reformed to the detriment of lithium's reputation as a highly effective stabilizer. As bipolar diagnosis evolved from a Kraepelinian clinical profile to DSM III to V symptom criteria [33], the incidence of bipolar diagnoses markedly increased, while lithium's efficacy became increasingly questioned [17,18]. By 2000, maintenance clinical trials emerged that included these loosely diagnosed bipolar disorders and found lithium not only ineffective but somewhat worse than placebo [13].

4. Reassessing Monitoring Requirements

Like many pharmacological substances, lithium can be toxic if used inappropriately. Therefore, out of caution, during the 1960s and 1970s, frequent determinations of lithium plasma concentration were required [34]. However, these guidelines were not sufficiently adjusted as clinical practice evolved.

Over time, Mogens Schou concluded that for long-term treatment, the critically important safety factor is sufficient patient education about the principles of lithium stabilization and compliance. He therefore developed and updated educational materials [35,36], explaining essential knowledge for patients and relatives regarding safe and effective lithium treatment.

Through six decades of treating with lithium, we have found that with proper medical assessment, correctly adjusted individual dosing, and sufficient patient education, blood value monitoring, including lithium levels every 6–12 months or when clinically indicated, is sufficient. This observation was made in lithium-responsive patients.

Lithium's potential toxicity is often misconstrued. The main risk emerges when lithium accumulates in tissues following prolonged intake or interacts with medical co-

https://doi.org/10.3390/ph19040527

medication. Otherwise, in the correct dosage provided to patients who are sufficiently medically screened, lithium is generally well tolerated.

Critical elements for safe lithium treatment include careful medical patient selection, correct determination of individual patient dosage, sufficient knowledge of pharmacological data, and patient education about the principles of lithium stabilization and medication compliance [37].

5. Rethinking Side Effects

Misunderstandings also exist regarding lithium treatment adverse effects. Lithium's adverse effects depend on numerous administration factors: formulation type (plain versus slow-release) [38], dosing amount and frequency [39], patient factors such as age and bipolar subtype, and variables known from general pharmacology.

Attention also needs to be paid to side effect differences between lithium stabilization responders and non-responders. Significant differences exist in adverse effects between these two groups. In a cohort of three hundred lithium maintenance responders, the most frequent complaint was increased thirst in one-third of patients, primarily at treatment onset [40]. Other lithium side effects were mild, often transitory, present in only 10% of patients, and often correctable. The only troublesome, disconcerting side effect was reduced glomerular filtration, becoming significant after decades of uninterrupted administration [8,14].

The above observation differs markedly from the adverse effects reported in undifferentiated lithium-treated populations [8], which fail to consider the above-mentioned factors: e.g., whether dosage was appropriate, whether lithium levels were therapeutic, whether lithium was properly indicated, or whether patients responded.

Virtually all observations have been collected and reported indiscriminately, combining all these different situations together. Elevated lithium side effects in such reports justifiably raise concerns [41], but they are not informative for properly run clinical practice. Included in these reports are the lithium side effects of patients who do not respond and should not continue with lithium maintenance, which lacks clinical utility. In my opinion, the particularly clinically informative has been published by Tondo et al. [42].

6. How to Reverse the Trend

Introducing long-term lithium treatment into clinical practice was challenging [1]. It is therefore worrisome to see the declining use of lithium in bipolar disorders, as it represents a significant loss for both patients and clinical practice. Yet this troubling development may be reversed through strategic interventions that would address the core obstacles that mistakenly limit lithium's utilization in practice. Such a path forward would require an approach encompassing improved patient selection for maintenance, innovative kidney protection strategies, and enhanced education across patients, families, and clinicians.

7. Selecting Lithium Responders in Advance

The first step in reversing this trend would rest in helping clinicians appreciate that a positive lithium maintenance response can be predicted with remarkable accuracy from each patient's clinical profile [43,44]. The ability to forecast lithium stabilization success emerged from five decades of systematic research, supported in particular by discriminant function [44] and machine learning (AI) analyses [31]. Utilizing this recognition markedly improves lithium treatment outcomes, reduces side effects, and simplifies clinical practice. The International Group for the Study of Lithium-Treated Patients (IGSLIs) continues to advance in this direction, currently designing an extensive collaborative study to incorporate established genetic correlates of response alongside clinical predictive features.

https://doi.org/10.3390/ph19040527

Before identifying lithium maintenance responders in advance in clinical practice, carefully designed prospective studies are needed.

Successful predictions from the patient's clinical profile depend primarily on the accurate evaluation of two factors: the type of clinical course and family history findings. However, these two predictors need to be defined in a distinctive way that is often overlooked in current clinical practice [44,45].

The most important predictor from the clinical course is that true "illness-free intervals" must be differentiated from "clinical recovery." An episodic bipolar course with psychopathology-free intervals requires a return to ordinary mental health. This differs substantially from "clinical recovery," where patients return to family and work responsibilities, often without complaining of any symptoms, yet closer clinical assessment reveals persistent fluctuations in anxiety, mood, sleep, cognition, and other residual symptoms. Only true free intervals predict a lasting lithium-stabilizing response.

To be predictive, family history assessment extends beyond information patients provide and needs to include reports from other family members, particularly a relative who possesses more comprehensive knowledge about psychiatric problems and mental health in the family. The identification of psychiatric disorders with episodic courses contributes significantly to an accurate prediction of lithium stabilization [46,47].

When the above assessment principles are applied, approximately one-third of patients who are currently broadly diagnosed as bipolar can be identified as having a lithium stabilization-responsive profile [48]. Some patients have bipolar propensity and lithium responsiveness, yet for a long time may present only with recurrent depression or recurrent disabling anxiety. The clinical implications of improved patient selection for lithium stabilization extend far beyond treating an individual patient. Once clinicians witness how identifying stabilization response in advance simplifies further clinical care and transforms patients' lives, their appreciation of lithium stabilization is likely to improve and may start reversing the downward trend in lithium use.

Observing the lasting stability in patients appropriately selected for lithium stabilization motivates clinicians to abandon the prevailing trial-and-error approach to mood stabilization and reduce excessive, unjustified polypharmacy that leads to frustrating outcomes and higher side effects. Treating bipolar patients who are lithium-stabilized and remain that way [49] liberates clinicians' time and resources for attending to the remaining two-thirds of bipolar spectrum patients who require different treatment strategies. Such experiences with lithium-stabilized patients also offer a clinical hint that lithium is usually a poor choice for successful maintenance treatment of other types of bipolar spectrum disorders.

It may be of interest that the patient's clinical profile, which was found to be linked with satisfactory stabilization response, tends to overlap with the description that Kraepelin provided for manic-depressive illness [15]. Also, the clinicians who, during the 1970s, completed successful double-blind lithium studies were diagnosing within the Kraepelinian tradition. The meta-analyses that now demonstrate the efficacy of lithium stabilization always carefully involve such studies [50], and they probably would not demonstrate lithium's efficacy without these studies.

8. Protecting Patients' Kidney Functions

Parallel to improving patient selection, protecting the patient's kidney function would represent another crucial step in preserving lithium treatment for bipolar patients. Schou et al. recommended continued daily lithium maintenance [51], as they were aware that the manic-depressive course is recurrent, that recurrences reappear capriciously, and that the cycle length is highly variable in individual patients. They advised lithium treatment

https://doi.org/10.3390/ph19040527

without any interruption because they worked mostly with severely ill patients who were finally stabilized on lithium for the first time in their lives, and the negative effect on kidney function was not known.

There was not much reason to worry then. In fact, after the first decade of lithium stabilization, an abnormally low glomerular filtration rate (referred to as GFR) was observed only in those individuals who showed abnormally low GFR during the medical screening for lithium treatment [52]. The GFR values started dropping substantially after some decades of continuous lithium intake.

Professor Rybakowski's team provided an important example through their unique report of five patients treated with lithium for over fifty years [14]. Despite receiving exemplary conservative lithium treatment with careful monitoring, patients developed GFR disruptions. I witnessed the quality of their care when I was allowed to review this bipolar cohort during my visit on behalf of an IGSLI collaborative study. Rybakowski's team's findings demonstrated again that lithium, regardless of expert management quality, possesses inherent nephrotoxic potential that manifests after decades of uninterrupted use.

We had similar observations in our Ontario bipolar patient cohorts. After the first decades of continuous lithium treatment, our lithium stabilization responders showed no demonstrable adverse effects on GFR [52] and only mildly increased urinary volume. The GFR decrease became evident after three decades of uninterrupted lithium treatment and exacerbated after four and five decades. Over one-fifth of our patients stabilized with lithium for more than four decades exhibited abnormally low GFR and, according to our nephrologists, attributable solely to lithium treatment [53].

However, learning more about the episodic course of classical manic-depressive illness opened new possibilities for kidney protection through intermittent treatment strategies. In the 1960s, with Professor Jules Angst, we gathered data from lifelong clinical courses of manic-depressive patients. At that time, it was still possible to obtain records from a large cohort of patients who were medicated only during the acute episodes and received no maintenance treatment [54]. Later, clinical course observations were added from bipolar patients free of mood stabilizers during their initial episodes.

For two decades, we then explored what may drive this seemingly capricious, recurrent process of bipolar illness. Finally, led by neuroscientists Anna Yamamoto and Josef Lat [55–58], we uncovered during the clinical courses largely predictable oscillatory patterns that neuroscientists had described in extensive explorations in animal brains.

The findings indicate that bipolar recurrences reflect brain oscillatory processes with excessive amplitudes poorly modulated. Clinicians easily miss this because observing natural, untreated courses of bipolar illness has now become a rarity. The brain is a huge assembly of oscillators active on all different neuronal levels. Because of the quality of neurons in the frontocentral areas [59], oscillations quickly expand and retract, and the activity above the usual ceiling may take place in episodes. In general, these oscillatory patterns are either linear or exponential, but otherwise individual.

The oscillatory nature of bipolar illness can be demonstrated retrospectively through sufficient clinical information or prospectively using serial dexamethasone suppression tests as brain arousal level proxies. This understanding enables the identification of individual high-risk and low-risk periods, creating opportunities for treatment interruptions during suitable periods. Intermittent treatment strategy retains lithium stabilization while it provides protection by reducing cumulative kidney exposure to lithium. Using this strategy, it is important not to underestimate the risk of recurrence reflected in the literature [60–62].

We observed that in patients with severe illness, recurrences develop during the peak of each cycle. In these patients, we first noticed the relationship between the oscillatory processes and recurrences. In patients with less severe classical bipolar illness, only the majority of the cycle peaks are associated with recurrences (presumably because of other biological and psychosocial factors at the time). However, the subsequent recurrences and neurobiological measurements indicated that in these less severe patients, their individual cycle lengths continue unchanged. In other words, the individual cycle pattern for each patient remained unchanged. Therefore, high- and low-risk time periods may be estimated and, in selected lithium responders, make intermittent treatment possible.

This lack of adaptability appears to be one of the characteristics of classical bipolar course; unless it is artificially sped up, e.g., by antidepressants. In comparison, in healthy subjects, the brain oscillatory patterns are adaptive, variable, and change under the circumstances [63]. More details of this approach are explained with examples in other publications [64] and in an upcoming book.

Successful and beneficial use of intermittent treatment for specific situations is not new in medicine and has been applied in chronic pain management, cardiology and psychiatry [65,66]. Our clinical experience with intermittent lithium treatment spans thirty-five years and has included thirty-one carefully selected patients during long-term lithium treatment [53].

Patients requesting intermittent lithium treatment had to meet a set of criteria to be considered suitable: a clearly documented fully episodic course; good response to continuous lithium maintenance for a sufficient time; the patient's significant other cooperating and consenting to the treatment plan; and feasible regular clinical supervision. These lithium-responsive patients typically requested a modification of their continuous lithium treatment to intermittent after they learned about lithium's potential nephrotoxicity. Therefore, these patients embody an extended clinical observation, not a representative sample. I need to stress the point mentioned above: these patients were all excellent responders to lithium maintenance.

The patients were placed on intermittent lithium treatment for clinical purposes, and no comparative study was then intended. Yet the preliminary findings in this pilot are promising for kidney protection. To obtain some quantitative indication, the age correlations of patients on intermittent treatment were compared with those of patients on continuous lithium and patients who never received lithium.

Preliminary analyses imply that in patients treated intermittently, GFR decrease with age resembles that of lithium-free controls. Thus, intermittent lithium treatment may avoid the excessive effects of uninterrupted, continuous long-term lithium treatment. The correlation between GFR decline and age in intermittent treatment patients ($r = 0.37$, $N = 31$) closely matches lithium-free controls ($r = 0.39$, $N = 100$), while continuous treatment patients indicate much higher GFR decline ($r = 0.68$, $N = 110$).

There is a widespread belief that discontinuing lithium medication is a high-risk step, often followed by a quick relapse, even with an intensity worse than ever before. This is true for bipolar spectrum disorders but does not apply to the classical type of bipolar illness. Here, the relapse comes only as one would expect from the previous course of the illness [27,28].

Over the decades of working with intermittent lithium treatment, bipolar depressive recurrences were observed in only three patients, all of them occurring when the patient stayed off lithium longer than they were advised, according to the duration of the low-risk period. Fortunately, in lithium stabilization responders, the lithium benefit can be promptly reproduced following discontinuation. After resuming their lithium dosage, all three patients fully recovered within 10–14 day [67].

These observations, while preliminary and obtained in specialized clinical settings, suggest that lithium's GFR effects might result from uninterrupted long-term exposure rather than from lithium itself. To test this possibility, other factors such as, for example,

https://doi.org/10.3390/ph19040527

cumulative lithium dose, need to be included in a future study. The pilot data presented here need to be considered hypothesis generating. For episodic bipolar illness, intermittent treatment offers hope for maintaining therapeutic benefits while avoiding the adverse effects of decades-long continuous exposure.

Alternative strategies for assessing recurrence risk are also being developed [68,69]. Furthermore, lowering the lithium dosage during low-risk periods rather than fully discontinuing also merits further exploration. If effective, some clinicians may prefer such less radical approaches to kidney protection. Moreover, in bipolar patients exhibiting inflammatory markers [70,71], it is important to consider adding an anti-inflammatory medication and reducing the lithium dosage to help protect the kidneys.

9. Improved Education

Finally, to reverse the trend, much needs to be done to educate psychiatrists, their patients, and family members about the principles of lithium stabilization and suicide prevention. It would be helpful to correct the prevailing beliefs and misconceptions about efficacy, side effects, and adequate monitoring of lithium stabilization.

10. Conclusions

Bipolar disorders afflict millions of individuals and are potentially lethal. Reversing lithium's declining clinical use through the outlined strategies would serve the broader goal of improving bipolar disorder treatment and reducing the great burden of these devastating illnesses on patients and their families.

When lithium is used appropriately in carefully selected patients with adequate protection and monitoring, it offers unique, unparalleled mood stability, suicide protection, and quality of life improvement. Restoring this treatment option through strategic interventions may represent a critical step toward optimizing care for all patients across the bipolar spectrum.

Funding: This clinical research received no external funding.

Institutional Review Board Statement: De-identified patient data were collected from several studies, all of which were approved by the respective Institutional Research Ethics Boards at McMaster University (1968–1988) and University of Ottawa, Royal Ottawa Hospital, University of Ottawa (1988–2023). The data were also included in previous publications and presentations [39,52].

Data Availability Statement: The clinical data are located in the Medical record Departments of Hamilton Psychiatric Hospital, Hamilton, and Royal Ottawa Hospital, Ottawa, ON, Canada.

Conflicts of Interest: The author declares no conflict of interest.

References

1. Schou, M.; Grof, P. History of lithium treatment. In *Lithium in Neuropsychiatry: The Comprehensive Guide*; Bauer, M., Grof, P., Mueller-Oerlinghausen, B., Eds.; Informa: London, UK, 2006; pp. 9–16.
2. Cloutier, M.; ExpertConsensus. Economic burden of declining lithium use in bipolar disorder. *J. Med. Econ.* **2022**, *25*, 234–245.
3. Greil, W. Twenty-three years of declining lithium use: Analysis of a pharmacoepidemiological dataset from German-speaking countries. *Pharmacopsychiatry* **2024**, *57*, 286–303. [CrossRef] [PubMed]
4. McKinney, J.R. Global trends in lithium utilization for bipolar disorder: A comprehensive analysis. *World J. Biol. Psychiatry* **2022**, *23*, 278–289.
5. Hidalgo-Mazzei, D.; Mantingh, T.; Pérez de Mendiola, X.; Samalin, L.; Undurraga, J.; Strejilevich, S.; Severus, E.; Bauer, M.; González-Pinto, A.; Nolen, W.; et al. Clinician attitudes and lithium prescribing in bipolar disorder: A systematic review. *J. Affect. Disord.* **2023**, *321*, 45–58.
6. Sleem, A.; El-Mallakh, R.S. Declining lithium use in mood disorders. *Int. J. Bipolar Disord.* **2021**, *9*, 15.
7. Olivetti, E.A.; Ceder, G.; Gaustad, G.G.; Fu, X. Lithium-ion battery supply chain considerations: Analysis of potential bottlenecks in critical metals. *Joule* **2017**, *1*, 229–243. [CrossRef]

https://doi.org/10.3390/ph19040527

8. Gitlin, M.J. Lithium side effects and toxicity: Prevalence and management strategies. *Int. J. Bipolar Disord.* **2016**, *4*, 27. [CrossRef]
9. Rybakowski, J.K. Challenging the negative perception of lithium and optimizing its long-term administration. *Front. Mol. Neurosci.* **2018**, *11*, 349. [CrossRef]
10. Vestergaard, P.; Schou, M.; Thomsen, K. Monitoring of patients in prophylactic lithium treatment. *Br. J. Psychiatry* **1982**, *140*, 185–287. [CrossRef]
11. Yildiz, A. Market dynamics and lithium decline: The role of second-generation antipsychotics in bipolar disorder treatment. *Bipolar Disord.* **2022**, *24*, 289–301.
12. Hestbech, J.; Hansen, H.E.; Amdisen, A.; Olsen, S. Chronic renal lesions following long-term treatment with lithium. *Kidney Int.* **1977**, *12*, 205–213. [CrossRef] [PubMed]
13. Bowden, C.; Calabrese, J.R.; McElroy, S.L. A randomized, placebo-controlled 12-month trial of divalproex and lithium in treatment of outpatients with bipolar I disorder. *Arch. Gen. Psychiatry* **2000**, *57*, 481–489. [CrossRef] [PubMed]
14. Ferensztajn-Rochowiak, E.; Chlopocka-Wozniak, M.; Rybakowski, J.K. Ultra-long-term lithium administration: All important matters and a case report of successful 50-year lithium treatment. *Braz. J. Psychiatry* **2021**, *43*, 407–413. [CrossRef]
15. Kraepelin, E. *Manic-Depressive Insanity and Paranoia*; Classics in Psychiatry; Arno Press: New York, NY, USA, 1976; 280p.
16. American Psychiatric Association. *Diagnostic and Statistical Manual of Mental Disorders DSM-5*; APA: Washington, DC, USA, 2013.
17. Deshauer, D.; Fergusson, D.; Duffy, A.; Albuquerque, J.; Grof, P. Re-evaluation of randomized control trials of lithium monotherapy: A cohort effect. *Bipolar Disord.* **2005**, *7*, 382–387. [CrossRef] [PubMed]
18. Deshauer, D. Evolution of bipolar disorder diagnosis and its impact on lithium efficacy studies. *J. Psychiatr. Res.* **2023**, *6*, 45–54.
19. Baastrup, P.C.; Poulsen, J.C.; Schou, M. Prophylactic lithium: Double blind discontinuation in manic-depressive and recurrent-depressive disorders. *Lancet* **1970**, *2*, 326. [CrossRef]
20. Baastrup, P.C.; Schou, M. Prophylactic lithium. *Lancet* **1968**, *1*, 1419–1422. [CrossRef]
21. Schou, M. The range of clinical uses of lithium. In *Lithium in Medical Practice*; Johnson, F.N., Ed.; University Park Press: Baltimore, MD, USA, 1978; pp. 21–40.
22. Garver, D.L.; Hirschowitz, J.; Fleishman, R.; Djuric, P.E. Lithium response and psychoses: A double-blind, placebo-controlled study. *Psychiatry Res.* **1984**, *12*, 57–68. [CrossRef]
23. Garver, D.L.; Hutchinson, L.J. Psychosis, lithium-induced antipsychotic response, and seasonality. *Psychiatry Res.* **1988**, *26*, 279–286. [CrossRef]
24. Hirschowitz, J.; Casper, R.; Garver, D.L.; Chang, S. Lithium response in good prognosis schizophrenia. *Am. J. Psychiatry* **1980**, *137*, 916–920. [CrossRef]
25. Perris, C. The concept of cycloid psychotic disorder. *Psychiatr. Dev.* **1988**, *1*, 37–56.
26. Grof, P. Has the effectiveness of lithium changed? Impact of the variety of lithium's effects. *Neuropsychopharmacology* **1998**, *19*, 183–188. [CrossRef] [PubMed]
27. Schou, M.; Thomsen, K.; Baastrup, P. Studies on the course of recurrent endogenous affective disorders. *Int. Pharmacopsychiatry* **1970**, *5*, 100–106. [CrossRef]
28. Grof, P.; Cakuls, P.; Dostal, T. Lithium drop-outs: A follow-up study of patients who discontinued prophylactic treatment. *Int. Pharmacopsychiatry* **1970**, *5*, 162–169. [CrossRef]
29. Muller-Oerlinghausen, B.; Grof, P.; Schou, M. Lithium and suicide prevention. *Br. J. Psychiatry* **1999**, *175*, 90–91. [CrossRef]
30. Tondo, L.; Baldessarini, R.J. History of suicide prevention with lithium treatment. *Pharmaceuticals* **2025**, *18*, 258. [CrossRef]
31. Nunes, A.; Ardau, R.; Berghöfer, A.; Bocchetta, A.; Chillotti, C.; Deiana, V.; Garnham, J.; Grof, E.; Hajek, T.; Manchia, M.; et al. Prediction of lithium response using clinical data. *Acta Psychiatr. Scand.* **2020**, *141*, 131–141. [CrossRef]
32. Post, R.M.; Leverich, G.S.; Altshuler, L.; Mikalauskas, K. Lithium-discontinuation-induced refractoriness: Preliminary observations. *Am. J. Psychiatry* **1992**, *149*, 1727–1729.
33. Severus, E.; Bauer, M. Diagnosing bipolar disorders in DSM-5. *Int. J. Bipolar Disord.* **2013**, *1*, 14. [CrossRef]
34. Schou, M.; Amdisen, A.; Baastrup, P.C. The practical management of lithium treatment. *Br. J. Hosp. Med.* **1971**, *6*, 53–60.
35. Schou, M. *Lithium Treatment of Mood Disorders, A Practical Guide*; Karger: Basel, Switzerland, 1980.
36. Schou, M. *Lithium Treatment of Mood Disorders: A Practical Guide*, 5th ed.; Karger: Berlin, Germany, 2004.
37. Grof, P.; Mueller-Oerlinghausen, B. Lithium monitoring in clinical practice: Evidence-based recommendations. *Int. J. Bipolar Disord.* **2023**, *11*, 8.
38. Grof, P.; MacCrimmon, D.; Saxena, B.; Daigle, L. Bioavailability and side effects of different lithium carbonate products. *Neuropsychobiology* **1976**, *2*, 313–323. [CrossRef] [PubMed]
39. Bowen, R.C.; Grof, P.; Grof, E. Less frequent lithium administration and lower urine volume. *Am. J. Psychiatry* **1991**, *148*, 189–192. [CrossRef] [PubMed]
40. Grof, P. Lithium Maintenance and Side-Effects in Responders. 2025; *unpublished data*.
41. Gitlin, M.J. Lithium-induced renal insufficiency. *J. Clin. Psychopharmacol.* **1993**, *13*, 276–279. [CrossRef]

https://doi.org/10.3390/ph19040527

42. Tondo, L.; Alda, M.; Bauer, M.; Bergink, V.; Grof, P.; Hajek, T.; Lewitka, U.; Licht, R.W.; Manchia, M.; Mueller-Oerlinghausen, B.; et al. Clinical use of lithium salts: Guide for users and prescribers. *Int. J. Bipolar Disord.* **2019**, *7*, 16. [CrossRef]
43. Grof, P. Responders to long-term lithium treatment. In *Lithium in Neuropsychiatry: The Comprehensive Guide*; Bauer, M., Grof, P., Mueller-Oerlinghausen, B., Eds.; Informa: London, UK, 2006; pp. 157–178.
44. Grof, P. Sixty years of lithium responders. *Neuropsychobiology* **2010**, *62*, 8–16. [CrossRef]
45. Grof, P. Response to long-term lithium treatment: Research studies and clinical implications. In *The Affective Disorders*; Davis, J.M., Maas, J.W., Eds.; American Psychiatric Press: Washington, DC, USA, 1983; pp. 357–366.
46. Grof, P.; Duffy, A.; Cavazzoni, P.; Grof, E.; Garnham, J.; MacDougall, M.; O'Donovan, C.; Alda, M. Is response to prophylactic lithium a familial trait? *J. Clin. Psychiatry* **2002**, *63*, 942–947. [CrossRef]
47. Duffy, A.; Grof, P. Longitudinal studies of bipolar patients and their families: Translating findings to advance individualized risk prediction, treatment, and research. *Int. J. Bipolar Disord.* **2024**, *12*, 12–20. [CrossRef]
48. Alda, M. Who are excellent lithium responders and why do they matter? *World Psychiatry* **2017**, *16*, 3. [CrossRef]
49. Berghofer, A.; Alda, M.; Adli, M.; Baethge, C.; Bauer, M.; Bschor, T.; Grof, P.; Muller-Oerlinghausen, B.; Rybakowski, J.K.; Suwalska, A.; et al. Stability of lithium treatment in bipolar disorder—Long-term follow-up of 346 patients. *Int. J. Bipolar Disord.* **2013**, *1*, 11. [CrossRef] [PubMed]
50. Cipriani, A. Lithium versus other mood stabilizers for bipolar disorder: Updated systematic review and network meta-analysis. *Lancet* **2023**, *401*, 501–512.
51. Schou, M.; Baastrup, P.C.; Grof, P.; Weis, P.; Angst, J. Pharmacological and clinical problems of lithium prophylaxis. *Br. J. Psychiatry* **1970**, *116*, 615–619. [CrossRef] [PubMed]
52. Grof, P.; MacCrimmon, D.J.; Smith, E.K.; Daigle, L.; Saxena, B.; Varma, R.; Grof, E.; Keitner, G.; Kenny, J. Long-term lithium treatment and the kidney. *Can. J. Psychiatry* **1980**, *25*, 535–544. [CrossRef]
53. Grof, P. Ultra-Long Lithium Maintenance and Kidney Glomerular Function. 2025; *unpublished data*.
54. Angst, J.; Grof, P. The course of monopolar depressions and bipolar psychoses. In *Lithium in Psychiatry, a Synopsis*; Université de Laval Press: Quebec City, QC, Canada, 1976; pp. 93–104.
55. Lat, J. The law of activation and psychosis. *Acta Nerv. Super.* **1978**, *20*, 34–35.
56. Lat, J.; Indrova, M.; Fischlova, V. The oscillatory course of psychotic and neurotic illness. *Activ. Nerv. Sup.* **1982**, *246*, 186–188.
57. Lat, J.; Yamamotova, A.; Indrova, M.; Muronova, M. The recurrent course of psychotic illness from point of view of activation and motivation dynamics. *Cesk. Psychiatr.* **1981**, *77*, 395–406.
58. Yamamotova, A.; Grof, P.; Wackermann, J. Results of chronobiological analysis of recurrent affective disorders. *Studia Psychol.* **1995**, *36*, 377–381.
59. Buzsaki, G. *Rhythms of the Brain*; Oxford University Press: Oxford, UK; New York, NY, USA, 2006.
60. Baldessarini, R.; Pinna, M.; Contu, M.; Vázquez, G.; Tondo, L. Risk factors for early recurrence after discontinuing lithium in bipolar disorder. *Bipolar Disord.* **2022**, *24*, 720–725. [CrossRef]
61. Baldessarini, R.; Tondo, L.; Hennen, J. Effects of lithium treatment and its discontinuation on suicidal behavior in bipolar manic-depressive disorders. *J. Clin. Psychiatry* **1999**, *60*, 77–84.
62. Suppes, T.; Baldessarini, R.; Faedda, G.; Tohen, M. Risk of recurrence following discontinuation of lithium treatment in bipolar disorder. *Arch. Gen. Psychiatry* **1991**, *48*, 1082–1088. [CrossRef] [PubMed]
63. Lat, J.; Yamamotova, A. The role of the time dimension in the analysis of spontaneous behaviour. *Physiol. Bohemoslov.* **1985**, *34*, 97–100.
64. Grof, P.; Ahrens, B.; Yamamotova, A.; Fox, D. Chronobiological nature of recurrence (In German: Chronobiologische Verlaufsmuster affektiver Erkrankungen). In *Ziele und Ergebnisse der Medikamentosen Prophylaxe Affektiver Psychosen*; Muller-Oerlinghausen, B., Berghoefer, A., Eds.; Thieme Verlag: Stuttgart, Germany, 1994; pp. 67–77.
65. Ananth, J. Intermittent lithium therapy in a bipolar patient. *Can. J. Psychiatry* **1990**, *35*, 344–345. [CrossRef]
66. Abe, K. Intermittent treatment of recurrent mood disorders. *Jpn. Psychiat.* **2006**, *49*, 18–21.
67. Grof, P. Intermittent Lithium Treatment. 2025; *to be published*.
68. Ortiz, A.; Halabi, R.; Alda, M.; Burgos, A.; DeShaw, A.; Gonzalez-Torres, C.; Husain, M.; O'Donovan, C.; Tolend, M.; Hintze, A.; et al. Day-to-day variability in sleep and activity predict the onset of a hypomanic episode in patients with bipolar disorder. *J. Affect. Disord.* **2025**, *374*, 75–83. [CrossRef]
69. Ortiz, A.; Halabi, R.; Alda, M.I.; Husain, M.; Nunes, A.; O'Donovan, C.; Patterson, R.; Mulsant, B.H.; Hintze, A. Day-to-day variability in activity levels detects transitions to depressive symptoms in bipolar disorder earlier than changes in sleep and mood. *Int. J. Bipolar Disord.* **2025**, *13*, 13. [CrossRef]

70. Chen, Z.; Huang, Y.; Wang, B. T cells: An emerging cast of roles in bipolar disorder. *Transl. Psychiatry* **2023**, *13*, 153. [CrossRef]
71. Poletti, S.; Mazza, M.; Benedetti, F. Inflammatory mediators in major depression and bipolar disorder. *Transl. Psychiatry* **2024**, *14*, 1. [CrossRef]

https://doi.org/10.3390/ph19040527